LEVERAGING
LINKEDIN

For
Job Search Success

FRED COON & SUSAN BARENS

GAFF Publishing

CONTENTS

Don't Make These LinkedIn Mistakes 153

Resources to Make Your Job Search on LinkedIn Easier 157

About the Authors 159

Sources 163

Index 165

COPYRIGHT

Getting Started with LinkedIn

Why Get LinkedIn?

Frequently, I hear that question. Not long ago, a friend asked me, "Why should I bother with LinkedIn?" He wasn't a technologically challenged individual, so I was surprised that he should ask. My answer was simple, "It is the top social networking website for job seekers."

As Jeff Weiner, CEO of LinkedIn, explains it, "Post a full profile and get connected to the people you trust. Because if you're connected to those people and you posted a profile, then when other people are searching for people, they might find you."

With more than 300+ million registered users—and adding two new members every second—the rate at which your network expands on LinkedIn can be truly amazing. A hundred strategic contacts could mean access to millions of people in a short amount of time. You'd have to attend dozens—or hundreds— of in-person networking events to equal the reach you can get on LinkedIn.

LinkedIn allows you to leverage the power of your network — the people you already know, and the people those people know — to help you connect to the person (or persons) who are in a position to offer you a job.

As the cofounder of LinkedIn, Reid Hoffman, puts it, LinkedIn is about "connecting talent with opportunity on a massive scale."

Executives from all Fortune 500 companies are on LinkedIn. In addition, 59 percent of folks who are active on social networking sites say LinkedIn is their platform of choice, according to a June 2011 report from Performics and ROI Research.

However, author Guy Kawasaki puts it best — "I could make the case that Facebook is for show, and LinkedIn is for dough."

Top 4 Things You Must Do on LinkedIn

If you don't do these four things, your LinkedIn profile won't deliver the same benefits others have experienced.

Complete Your Profile. Your profile is the front door to your LinkedIn account. First impressions matter, so make sure you've made your profile as complete as possible. Then make sure you keep your profile up to date, accurate, and complete.

Grow Your Connections. There are two schools of thought when it comes to LinkedIn connections. You can choose to connect selectively, accepting invitations only from those you know and trust, or you can use LinkedIn to grow the network of people you know. You can connect with people you meet through Groups and get introduced to people you don't know offline.

Give to Get. Authentic, genuine recommendations can make or break a LinkedIn profile (just like references can for

POWER TIP

Make sure your recommendations are specific and detailed. You don't want a recommendation that sounds like it could fit anyone. Quantify accomplishments (with percentages, numbers, and dollar amounts) as much as possible.

POWER TIP

The power of networking lies in friends of friends, so the larger your network, the easier it will be to connect with someone you don't know (yet). Remember the principle of six degrees of separation.

a job candidate). Instead of sending out those presumptuous LinkedIn "Can you endorse me?" emails, select a handful of people in your network and write recommendations for them without asking for one in return. You will be surprised at how many people will reciprocate.

Get Involved. Join some LinkedIn Groups. Groups are the water cooler of the social site. You can find Groups for school and university alumni, your former and current employers, trade groups, industry associations, and more.

Top 7 Reasons to Be on LinkedIn

1 Because you find business professionals there

Other social networks may try to focus on attracting professionals, but none compare with LinkedIn. Most are niche oriented.

LinkedIn is not! Maybe that's why its membership has grown to over 300 million members since its launch in May 2003. No job seeker can afford to ignore its power, especially if the target job is with a younger company that is using social media in its marketing mix. People you know are already on the site, and so are people you should get to know — recruiters and hiring managers.

2 Because you need to "dig your well before you're thirsty"

In his book of the same name, author Harvey Mackay advocates building your network before you need it.

If you are presently employed, join LinkedIn now. That way you already have a network of connections in place when your job search begins.

3 **Because it never hurts to strengthen your offline network**

We've all lost track of people over time. Often, LinkedIn becomes a place you can reconnect. You find out what they are doing, where they work now, and whom they know—something that can be very important for a successful job search.

4 **Because reconnecting with former coworkers can lead to work**

Staying in contact with former coworkers can be difficult. You might not be the only one who has had to move around! LinkedIn makes reconnecting easy in two ways. You can search by name and by employer. Now, that reference that you really wanted might be as close as a reconnection.

5 **Because you can establish yourself as an expert**

Increasing your visibility is one of the ways you can position yourself as a thought leader in your industry. LinkedIn gives you a place to participate in Groups related to your expertise. LinkedIn's Answers forum is another opportunity to demonstrate your knowledge. Posting in Groups or answering questions, automatically adds content to your profile. When you actively engage in Groups on LinkedIn, visitors to your profile can see what you know and get a picture of how valuable your contribution could be to their business.

6 Because you may be found for jobs when you aren't even looking

Being asked to interview for a job based on your LinkedIn profile does happen. As a passive candidate, your robust LinkedIn profile—filled with your accomplishments and strong keywords—often leads prospective employers to you. More and more recruiters are searching LinkedIn to find candidates that match their search assignments.

7 Because your LinkedIn presence helps when someone "googles" you

If you Google a friend of mine, Denise Rutledge, you'll discover that she has a lot of competition—from a singer to a very nonbusiness-like person. Because she chose her LinkedIn ID before any other Denise Rutledge optimized their account, she is the first Denise Rutledge (at the time of this writing) to appear on Google's LinkedIn results. And that LinkedIn result is on the first page in search results.

The point is that hiring managers and recruiters usually Google their job candidates. If you have an optimized LinkedIn profile, your chances of appearing on the first page in Google is high.

Ultimately, a LinkedIn profile is a resume, business card, and elevator speech condensed in one place. It's a powerful marketing tool—one you can't really afford to ignore.

Why LinkedIn Is Important for Your Job Search

CNN Money reported just a few years ago that Accenture, a giant consulting firm, was going to forego the traditional methods of hiring headhunters or asking for employee referrals. The company even decided to avoid the job boards. It went straight to LinkedIn.

Just suppose you had been a telecom consultant and didn't have a LinkedIn presence? You would not have been found for one of the 50,000 job openings Accenture had to fill.

This is a trend, not a fad. You can expect in the future that many businesses will do most of their own head hunting. And they are going to start on LinkedIn!

Once upon a time, attending networking mixers, industry events, and Chamber of Commerce meetings were the best way to make new connections and build business relationships. Now, these activities are online within the LinkedIn community. Much like networking in person, professionals interact on LinkedIn with the explicit intention of making business connections.

With LinkedIn, you get all the benefits of networking in person, with less of the hassle. Instead of going from business lunch to business lunch hoping to meet people, LinkedIn provides a platform for you to specifically search and research individuals whom you know will directly add value to your job search.

We've already mentioned how employers and recruiters are using LinkedIn to locate both active job seekers and those who aren't necessarily looking (passive candidates).

On Jobfully Blog, the author reports that a friend was invited to interview for a job with a major mobile company. She wasn't even looking for a new job. The company just liked her LinkedIn profile. Several interviews later, she got the offer.

LinkedIn has more benefits. You have the ability to identify, research, contact, follow up, engage, and maintain your contacts in one place. In a world where information overload is a constant threat, that's a powerful organizational tool. No other platform has LinkedIn's ability to facilitate business networking. Facebook is for fun. LinkedIn is for business.

POWER TIP

Ensure your resume and LinkedIn profile are always in sync, as prospective employers are likely to Google you and compare the two.

Since LinkedIn is a one-stop snapshot of your background, some suggest that your LinkedIn profile is more important than your resume.

At the same time, you should always recognize that your LinkedIn profile is not your resume. LinkedIn is a personal branding page. You need both a resume and a LinkedIn profile. When in a job search there are two types of employers. Those who are using the latest technology, including social media, and those employers who prefer to post job openings in traditional ways and then use social media such as LinkedIn to narrow down the field of applicants that appear promising.

Keep your resume and LinkedIn profile in sync with one another, but they should not be exact copies of each other. Your resume and LinkedIn profile should agree on the positions you've held, your educational credentials, professional memberships, etc., but the content you include on your LinkedIn profile will be different from what you share on your resume.

For one thing, a resume is limited in what it can share about you. Moreover, it should be tailored to a specific job. LinkedIn offers far more creativity in marketing yourself. We're going to share strategies you can use on LinkedIn to find a job. But first, we need to cover the basics of setting up a LinkedIn account.

How to Set Up a Basic Account

Get Started

Setting up a LinkedIn account is a quick and easy process. However, speed is not the objective if you want to use LinkedIn to facilitate your job search. Rushing could lead to a sloppy profile that doesn't represent you well—or may even prevent you from being called for an interview.

Basic memberships in LinkedIn are free. For most job seekers, the free option is adequate to effectively network on the site. (If you find you need the paid functionality, you can always upgrade your account later.) To get started: Go to LinkedIn.com. Fill in your first and last name, email address, and password. Then click **Join Now.**

You will see an expanded Industry tab. The Industry that comes the closest to the service of Mr. Wordsmith is "Writing and Editing," You may be uncertain as to which job title or which industry you should choose. Pick the one that seems the most appropriate. You can always change the information later.

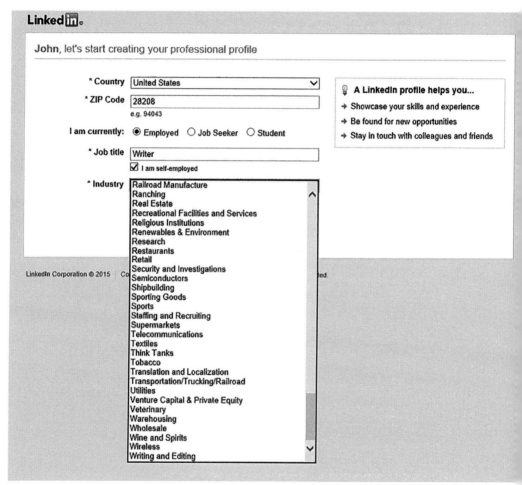

Source: **https://press.linkedin.com/about**

Worldwide Membership

- LinkedIn operates the world's largest professional network on the Internet with more than 347 million members in over 200 countries and territories.

- Professionals are signing up to join LinkedIn at a rate of more than two new members per second.

- In Q4 2014, more than 75% of new members came to LinkedIn from outside the United States.

- There are over 39 million students and recent college graduates on LinkedIn. They are LinkedIn's fastest-growing demographic.

347,000,000+
REGISTERED MEMBERS

Next, you'll see a screen to help you get started in building your profile.

Unlike other social media sites, it's not enough to just enter your name and email address to create a profile that you can complete later.

LinkedIn requires you to input details related to your career right away.

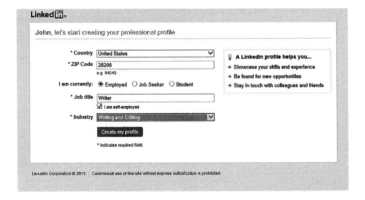

LinkedIn will then ask for your email address.

If you own a business and have your email tied to your website, then you probably do not have a web-based email account such as Gmail, Yahoo, Hotmail, etc.

LinkedIn will give you the warning message below. Simply obtain a web-based email account or select **Skip this step**.

This action is LinkedIn's attempt to import your email's address book. For now, we will **Skip this step**.

When you skip this step, you will receive the pop-up window below. Click **Skip**. You can always import your contacts later.

Next, LinkedIn will ask you to "Confirm your email address." Go to your email and click the link inside the email to confirm your new LinkedIn account.

The link will take you to the LinkedIn login page and ask you to log in. Once you successfully log in, you will see the screen below.

STRATEGY TIP

Use your personal email address when you join LinkedIn not your work one. If you ever change your current employer, you will lose access to your LinkedIn account when your employer terminates your email account. You don't want to lose all of your hard work.

LinkedIn will now take you through a systematic process to build your profile. In reality, these first steps are about data mining. LinkedIn is attempting to map you to others using your address book and other personal information.

We will control the building of your profile after we skip these steps. Again, you can add your contacts and control your connections later.

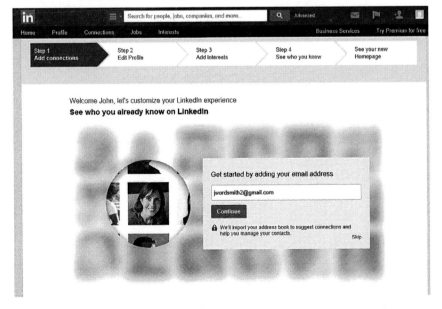

In Step 2. LinkedIn will ask you about your education. For now, click **Skip**.

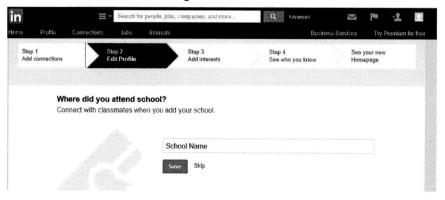

LinkedIn will ask you to "Add Interests" in Step 3. Feel free to follow a person that you find interesting. Following a person simply means you will automatically receive their notifications (e.g., thoughts, articles, etc.). You do not have to follow anyone. This is your preference. After you make your selections (or not), click **Continue.**

For Step 4, LinkedIn may show you photos of anyone you know and ask if you would like to connect with them. We will skip this for now and move to the next step. You can always add them later.

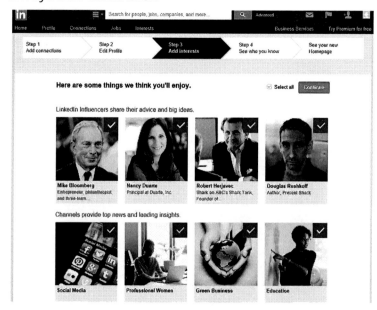

LinkedIn will then ask you to select an account that works best for you. To understand the different types of accounts, click **Compare accounts** in the lower left-hand corner.

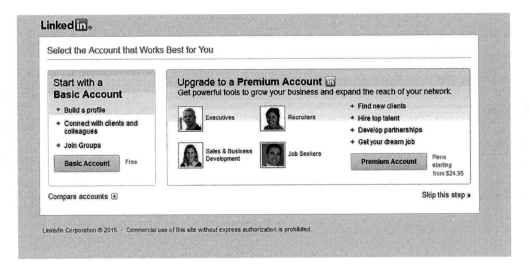

The "Basic Account" (Free) works fine for most people just starting out. Most functions that are available with a premium membership won't be important to you at this stage.

After you become familiar with LinkedIn, you can always upgrade later.

For now, select "Choose Basic," and LinkedIn will take you to your new homepage, as shown on the next page.

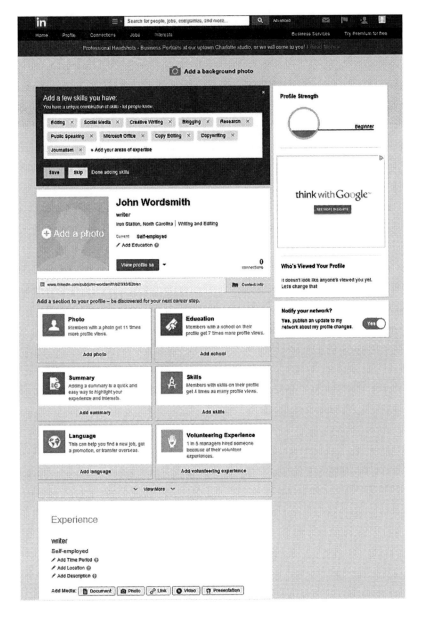

The next lesson will help you build your new profile. Although not required, we recommend that you bookmark this page for easy access. Otherwise, you can simply log in again (make sure your login name and password are handy).

In the next few lessons, you will learn how to create a backdoor link to your profile.

LinkedIn may ask you to confirm and review your email addresses affiliated with your new account.

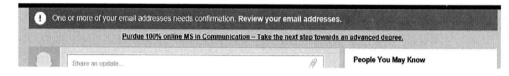

LinkedIn recommends that you add one personal address and a work address.

However, be careful when adding work email addresses because if you change employers, you will no longer have access to that box when your employer terminates that email account.

After adding another email, go to that email to confirm it with LinkedIn. Once you click on the link in the email, you are finished.

Take Action!

The next lesson will walk you through the process of building your LinkedIn profile. To prepare, you may want to gather the most important information about yourself. A resume is a handy reference tool when it comes to adding your employer, title, and dates. You will also need a quality, professional photograph.

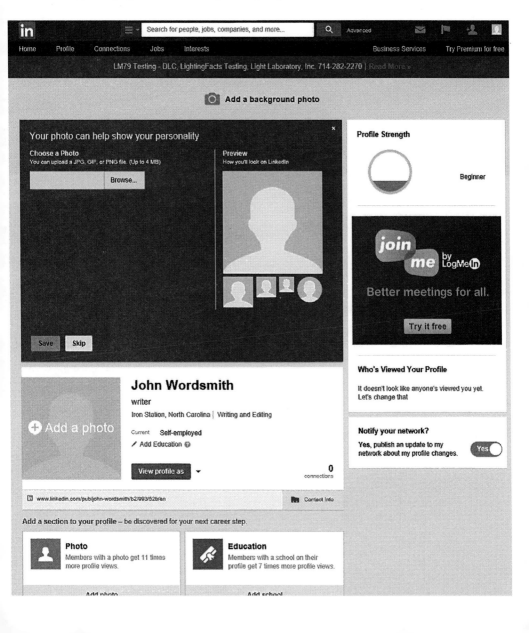

OPTIMIZING YOUR LINKEDIN PROFILE

Your LinkedIn home page includes a graphic that tracks your profile strength, encouraging you to add positions and education, include a summary and a photo, and ask for a recommendation in order to complete your profile. Here's how to find the page that shows your profile strength's status. Click the **Profile** tab on your main screen.

This is another way to get back to a page that asks questions and automatically populates the information. You'll also notice a box at the right that tells you your profile strength from LinkedIn's perspective.

According to LinkedIn, the strength and completion level of your profile determines a member's All-Star Status. The Profile Strength meter (visible to the right of your Profile page) will rise as you add more content. Profiles that are 100 percent complete rank higher in search results versus those who have less complete profiles.

Profile Strength

Beginner

A well-written and complete LinkedIn profile is essential to maximizing LinkedIn's application to your job search efforts. The structure of a LinkedIn profile is like a traditional resume. LinkedIn treats the words you use to describe your experience and education as keywords. Applying the correct keywords will help your profile rank higher in search results.

Importing Your Resume to Populate a Profile

LinkedIn offers members the option to import their resume information to populate the experience or education sections of your profile. To import your resume:

1 Click the **Profile** tab at the top of your homepage.

2 Next to the **View Profile As** button, you will see a drop-down arrow (▾). Place your cursor on the arrow and click **Import Resume.**

3 To locate the resume on your computer, click **Choose file** or **Browse.**

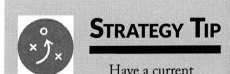

STRATEGY TIP

Have a current resume at your side to make sure you get dates and positions entered accurately. LinkedIr organizes your information based on the order in which you enter it, so work from the most recent position in reverse chronological order. It will save you valuable time.

4 Click **Upload** resume and LinkedIn will give you the option to review the information it extracted from the resume.

If the information is correct, click **Save.**

Making Connections

In addition to your profile strength, your connections affect your LinkedIn search results. While it is important to make many connections, you don't want to spam people. (Don't connect with people just for the sake of having significant connections. That's like giving your business card to everyone at a party.)

Look at other LinkedIn profiles of people who have your job title. See what they're including in their profile that gets them a high ranking using LinkedIn's search algorithm.

Find Career Profiles Using Google

How can you find similar LinkedIn profiles of individuals who do what you do?

Use a Google search as a fast way to retrieve similar profiles.

1 Go to Google.com.

2 Type in this search string: "Site:LinkedIn.com [your job title]."

3 Replace "your job title" with your job title or keywords or terms related to your job or industry.

Ignore the ads and focus initially on the top 10 search results. Some of the links will lead to individuals and others to related categories of professionals. You can learn a lot by visiting the top-listed profiles. Pay particular attention to the headline the professional has used to describe him or herself. Take notes on the keywords they are using. Also, you don't have to reinvent the wheel—if you have invested in a well-written professional resume, you will have most of the content you need to create a compelling LinkedIn profile.

Similar to other social media sites, LinkedIn uses its search algorithm to help connect you with people you know—or should know. For this reason, optimizing your profile so it ranks well in search results will help others find you.

Edit/Enhance Your Profile

Now that you have a better idea of what you want to include in your profile, it's time to start adding the information. There are several ways to create a profile.

First, LinkedIn will give you the option to build your profile as you wish. Simply select the section you want to populate (see the image on the right), and LinkedIn will take you through a series of questions.

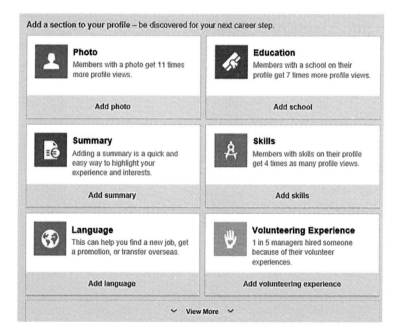

Another option is to fill out the blue box (as shown) that will appear at the top of your profile page each time you log in to LinkedIn. Even after you have an All-Star profile, this blue box will still appear.

The purpose is to remind you to keep adding different sections of your profile that may not be complete or to prompt you to add new skills that you may acquire as your career grows.

The last option is to go directly into edit mode. To do this, you can either go to the top of your homepage and hover your mouse over the **Profile tab**. Click **Edit Profile** and scroll to the areas you would like to edit content.

Clicking **Edit Profile** will take you to the window that allows you to modify any information you have already entered. You'll be visiting

this page frequently as you improve your profile!

A second option is to scroll to the area of your profile that you would like to edit. As shown in the next image, you will see a pencil icon next to the content. Click the pencil icon and LinkedIn will take you directly into edit mode. After you make additions, deletions, or edits, please be sure to click Save, otherwise your work will be lost.

How to Have a Profile that Stands Out

Standing out with your LinkedIn profile can mean highlighting the strongest qualifications you have for an employer in your LinkedIn headline, backing up those qualifications with accomplishments throughout your profile summary, and using strategies that will help you be found by the people who most need someone like you.

Don't try to be all things to all people. Although you can create different versions of your resume to target different types of positions, you're limited to one LinkedIn profile. On LinkedIn—as on your resume—one size does not fit all.

The most difficult part of creating your LinkedIn profile is sounding original. By articulating what makes you unique and valuable, you will attract the attention of prospective employers. Be specific about what distinguishes you from others with a similar job title.

The answers to these questions may give you some ideas for creating your LinkedIn profile and headline:

- *What specific job titles are used to describe someone in your position? (Be specific regarding level, function role, and industry.)*

- *In performance reviews, in what areas do you receive the highest scores or the most positive feedback?*

- *What is the most important part of your current job?*

- *What is your biggest achievement in your job—have you saved your company money, helped the company make money, or helped it become more efficient, improve safety, improve customer service, etc.*

- *What are your top three skills?*

- *What are you best known for at work?*

- *If you were asked to select your replacement, what qualities would you be looking for?*

- *What kind of challenges at work do you most enjoy working on?*

- *Do you have any specific training or credentials that distinguish you?*

- *What makes you different from other (job titles)? Is there an area in which you are better than others?*

Can you distinguish yourself by the geographic area in which you work or through your years of experience?

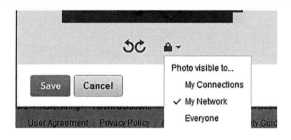

Hiring managers tend to make initial decisions in 5 to 7 seconds. So job seekers must establish "visual connectivity" in that amount of time. Strong profile pictures help to achieve this!

                     --**Jay Block**, President, The Jay Block Companies, LLC.

Jay's 6 Tips to Ensure the Right LinkedIn and Online Profile Picture

1   **Hire a professional.** A professional photographer can capture the essence of what job seekers want to portray. For a few hundred dollars, the return on investment can be invaluable.

2   **Choose a photo that looks like the job seeker.** When a job seeker appears at the interview, the interviewer should not be asking, "Where is your daughter?"

3   **The face ideally should take up about 60 percent of the frame.** Anyone can look good with the right photographer. A long distance picture of a job seeker on top of a mountain won't cut it.

4   **Choose the right expression.** Tyra Banks would say, "Smile with your eyes." Physiology is important when taking a picture. The picture should be energetic and engaging.

5   **Job seekers should wear what they would wear to work** (LinkedIn and websites). They should dress appropriately, even if the picture is simply a head and shoulder shot. They should wear clothes that match the level of attire for the positions they are seeking.

6   **Stand out professionally.** Again, a professional photographer can help here; black & white or color? What kind of background? Stand on your feet or on your head? What must one do to stand out professionally?

A picture tells 10,000 words. Make each word count.

Upload a Picture

If you skipped adding a picture earlier, you can attach your photo by choosing Profile from the navigation toolbar. Then click **Edit Profile** from the drop-down menu. Once on the editing page, choose **Add Photo**. (If you are

still building your profile, it may appear as **Add a picture** under the **Profile Completion Tips** section.)

Click the **Browse** button to find the photo on your hard drive you want to use. Then click **Upload Photo**. LinkedIn provides a built-in photo-cropping feature to capture your headshot. You can upload a JPG, GIF, or PNG file, but it should be square and cannot exceed 10MB. Ideally, it should be 400 x 400 pixels.

After saving your photo, you will have the option to designate who will see your image. Below your picture, you will see an image of a padlock with a pull-down menu (see screenshot below) next to it. You can choose to have it viewable by **My Connections, My Network**, or **Everyone.** For maximum expo-

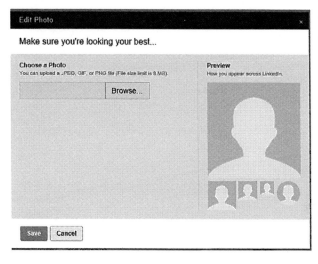

sure in your job search, choose Everyone. Depending on your comfort level and privacy concerns, you may want to limit the visibility to My Connections or My Network.

Create Your Headline

Your LinkedIn headline is the most important part of your profile. How you describe yourself to prospective employers and networking contacts is vitally important.

When someone conducts a search on LinkedIn, a search box returns a listing displaying only photos, names, and headlines. This is why it's important to have a good headline. A headline filled with the right keywords is an effective positioning tool. If you've selected your keywords properly, you will show up at the top by default.

If you add a new employer that you designate as "current," LinkedIn updates your headline automatically to reflect the new employer. For example, John Wordsmith is a staff writer at a Fortune 500 company, and LinkedIn automatically reflects this in his headline (under his name), as well as his current employer underneath the city and state.

Editing/Branding Your Headline

Today, branding statements are critical. What is your brand? Why are you unique?

Customizing your headline (under your name) to read as a personal branding statement is optional, but recommended. Simply go to **Edit Profile** and click the pencil icon next to your photo. Type your unique branding statement in the box and click **Save**.

The character limit is 120 characters, so writing a focused headline is important. It's good to include keywords in your headline, but don't limit yourself to keywords because it will look a bit choppy.

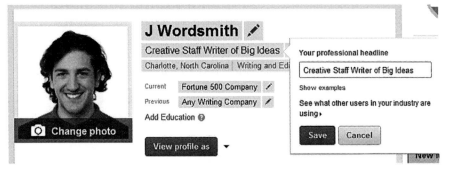

Writing Attention-Getting Headlines

The headline and the first two to three sentences of your LinkedIn profile summary are critical to making connections and securing opportunities with recruiters and hiring managers.

Instead of focusing on yourself, focus on what you offer a prospective employer. The information you provide should be 80 percent about what you have done for your current employer (accomplishments-oriented) and 20 percent about you and what you're looking for. Unfortunately, most LinkedIn profiles (especially the summary section!) are the reverse.

Think of it this way: Prospective employers tune into a particular radio station—it's called "WIIFM." All employers are listening for "What's In It For Me?" (WIIFM). *Remember: Employers hire for their reasons, not yours.*

What proof do you have that you can offer the employer

the results they are seeking? Quantify your accomplishments as much as possible in terms of numbers, percentages, and dollar amounts. Don't copy someone else's LinkedIn profile. Be original! Look at other profiles for ideas, but don't copy someone else's headline or summary. Remember—your online presence must speak to your uniqueness. Also, give your profile some personality!

People who make a connection with you through your profile are more likely to contact you about a career opportunity.

There are generally two schools of thought when it comes to writing your profile headline. The first is use a narrative or descriptive title; the second is to simply use keywords separated by commas, bullets, or the pipe symbol (|) on your keyboard.

LinkedIn's current algorithm gives higher ranking to matching keywords, so strategy number two appeals more to computer searches, while strategy number one appeals to human readers. Eventually, a human being will review all profiles found through computer searches. It is important to balance readability with the inclusion of keywords.

You are limited to just 120 characters in your LinkedIn headline, so it's also important to be succinct and direct.

Things you can consider including in your LinkedIn headline:

- Job titles

- Types of customers/projects you work with

- Industry Specialization

- Brands you've worked for

- Certifications or designations

- Geographic territory specialization

To improve readability, experiment with capitalizing the first letter of each of the words in your headline.

Formulas for Writing an Effective LinkedIn Headline

One effective technique is to pull a quote from a reference you've received plus another important piece of information such as a specialty area. Here are some strategies for writing your LinkedIn headline, along with the advantages and disadvantages that go along with each tactic.

1 Simple

Say it simply and directly. Give your job title and the name of your employer. This is a good strategy if your job title is a strong keyword and/or the company you work for is well known.

The advantage is that it clearly communicates who you are and what you do. The disadvantage is that it doesn't set you apart from any others who could claim those same credentials.

J Wordsmith
Staff Writer at Fortune 500 Company
Charlotte, North Carolina | Writing and Editing

Current　Fortune 500 Company
Previous　Any Writing Company
Add Education ⊕

View profile as ▼

0
connections

www.linkedin.com/pub/j-wordsmith/b2/994/1ba/en　Contact Info

This strategy can also use the following formulas:

Job Title

- Job Title at Company Name

- Job Title for Industry at Company Name

- **Job Title** Specializing in **Keywords**

Here is an example of a headline that incorporates a job title and keywords:

J Wordsmith

Creative Staff Writer of Big Healthcare Ideas

Charlotte, North Carolina | Writing and Editing

Current	Fortune 500 Company
Previous	Any Writing Company
Education	The George Washington University

View profile as ▾

0
connections

2 What You Do

This strategy focuses on job functions instead of job titles. The advantage to this headline strategy is that job functions often make excellent keywords. The possible problem is you simply string together a bunch of job functions without creating a story to explain who you are—along with what you do—so make sure you add some context to your keywords/job functions.

This strategy can also incorporate key projects and/or the names of key clients or important employers, especially if any of those have high name-recognition value. You may also wish to include a specific industry or geographic area to your job function-focused headline.

Here is an example that uses job function and targets the kind of clients this professional serves:

J Wordsmith

Creative Writer of Technical & Healthcare Trends

Charlotte, North Carolina | Writing and Editing

Current   Fortune 500 Company

Previous   Any Writing Company

Education   The George Washington University

View profile as ▾   con

3 The Big Benefit

It's important to identify the primary benefit you have to offer a prospective employer. Target what author Susan Britton Whitcomb says are "Employer Buying Motivators" in her book Resume Magic.

The 12 specific needs a company has include the company's desire to make money, save money, save time, make work easier, solve a specific problem, be more competitive, build relationships or an image, expand their business, attract new customers, and/or retain existing customers.

How can you be a problem solver for your next employer? Think about the job you want and what your next boss would want in an employee. Make that the focus of your headline.

This can be expressed in several different ways:

- (Job Title) That Gets (Results)

- (Adjective) (Job Title) With Recorded Success in (Results)

Be specific! Adding numbers and other specific wording can make your LinkedIn headline much more powerful. Here is the same strategy, but this one quantifies the scope and scale of the benefit to the employer.

Here are two examples

Fred Jobseeker

Secure Sponsor Support and Fan Attention for
International Professional Cycling Teams

Phoenix, Arizona | Public Relations and Communications

Fred Jobseeker

Secured Multi-Million-Dollar Sponsor Support along
with 12 Million+ Fan Impressions for International Pro
Cycling Team

Phoenix, Arizona | Public Relations and Communications

As you write your headlines, try not to include any of the
Top 10 Overused Buzzwords in LinkedIn Profiles in the
United States. Demonstrate that you truly are creative by
finding a way to show it. If you want to highlight your creativity, find a way to show it. Saying you're creative, innovative, etc. wastes valuable real estate on your profile.

POWER USER TIP — BE UNIQUE

According to LinkedIn, these are the 10 most overused words/
phrases on the site. Avoid using them in your headline and summary:

✓ Extensive Experience	✓ Proven Track Record
✓ Innovative	✓ Team Player
✓ Motivated	✓ Fast-Paced
✓ Results-oriented	✓ Problem Solver
✓ Dynamic	✓ Entrepreneurial

4 An Enthusiastic Testimonial

This headline strategy works best when you've received
honors or recognition within your field. This can be

an extremely effective strategy if you word it correctly. It's important that the designation is clear enough to stand on its own without too much detail. If it requires too much explanation, you may not have enough room within LinkedIn's 120-character limit.

A word of caution, however, don't trade on honors or recognition that are too far in the past. "Four-Time President's Award-Winner for Revenue Growth in the Ball Bearings Industry" isn't as impressive if those awards were for 1998, 2001, 2003, and 2005.

Susan B.

"Susan is a difference maker and realist who brings out the best in clients."

Charlotte, North Carolina Area
Public Relations and Communications

This strategy also works if you can make a claim that is defensible (i.e., the statement is arguably true). Put the claim in quotes so it appears as if it has been published somewhere.

If you are having trouble writing your LinkedIn headline, write a very rough draft. It doesn't matter if it's not good or if you have to leave some blanks. Having a framework will make it easier for you to complete later.

Go ahead and finish writing the rest of your LinkedIn profile and then come back to it. Oftentimes, the headline will become much clearer at that point. (Just remember to review your LinkedIn profile to make sure all the information you've included supports the focus of the content, as directed by the headline and summary.)

You can also look on LinkedIn for inspiration. Check out the headlines and summaries of people you're connected with, or conduct a search for others in your field. Just remember not to copy their information; instead, use it as inspiration.

5 Years of Experience

Another strategy is to highlight the number of years you've been in your industry.

You could also tweak LinkedIn's default heading by placing the industry you work in between your job title and

Susan B.

"Susan is a difference maker and realist who brings out the best in clients."

Charlotte, North Carolina Area
Public Relations and Communications

Current   Stewart, Cooper & Coon

the company name.

When appropriate, you might want to highlight a benefit you've consistently delivered. It's best to pick something that you know will show up in references.

When you are through with your headline, it should accomplish four things:

Susan B.

Executive Career Development & Coaching | Strategic Analysis & Communications - "Intelligent CareerCom Solutions"

Charlotte, North Carolina Area
| Public Relations and Communications

- It should be as powerful as a newspaper headline.

- It should be specific.

- It should attract the type of interest you want to draw.

- It should be clear what your skill set is.

LinkedIn Success Stories

John actively maintained his LinkedIn profile. He received two to three invitations a week to consider openings. One of those openings was a position at IBM. He interviewed and was offered the job.

David went to networking events and followed up on LinkedIn with all the contacts he made using Cardmunch, a LinkedIn application for iPhones that turns business cards into contacts. When he was laid off, he updated his LinkedIn status with "I'm up for grabs. Who wants me?" He received a phone call that same morning with a job offer. Within a week, he was back to work. Later when he was ready for a new opportunity, he was able to once again tap into his LinkedIn network. Within two weeks, he had found the opportunity that fit him.[1]

One of Kristen Jacoway's clients tailored his LinkedIn Groups search to the city and state in which he lived. He found out that some of the groups held live meet-ups. The HR specialist of his dream company was present at one of the meet-ups he attended. He had his resume with him and within a week he had an interview and was hired.

When Kristen Jacoway moved to South Carolina, she

---

1   From Kristen Jacoway interview, "Helping Clients with 'I'm in a Job Search—Now What?'" Jacoway is the author of I'm in a Job Search—Now What?

didn't know anyone. She found a company she was interested in working for. The head of HR was a 2nd degree connection. She clicked on her name and discovered which one of her 1st degree connections could introduce her. Within a week she had an interview.

A week later she was hired to provide contract work for the company.

3 Tips from Kathy Sweeney

1 Sign up for your LinkedIn account using a personal email, not a business one. That way you own your contacts and your profile, not your employer. If you use a company email address and you are fired or downsized, your email address will be deleted, rendering you unable to access your LI account and/or receive email updates.

2 Continually download your contacts into the email program you use. All you have to do is go to your Contacts pages in LinkedIn, look in the lower-right hand corner for the Export Contacts button, and proceed from there.

3 Do NOT check the box that says you are "looking for career opportunities" unless you are unemployed. Sweeney has 15 different executive recruiters who refer candidates to her to write their resumes. She has discussed their sourcing process regarding LinkedIn at length with each of them.

When they are looking for candidates, they could care less whether someone has "looking for career opportunities" marked in their profile. The recruiters are going to contact candidates anyway if they possess the qualifications that recruiters are trying to match up with job-order requirements.

Learn more about Kathy Sweeney at her website:

http://www.awriteresume.com/KathySweeneyNRWAVP.htm.

Additional LinkedIn Headline Tips

You got a peek at how to do this earlier, but we'll share a few more tips that will make the technical side easier to deal with for the nontechnical person.

Once you sign into your account, mouse over **Profile** in the Navigation bar. From the drop-down menu, click **Edit Profile.**

Click the blue pencil that shows up beside your headline in the new window. This will open a window. You'll notice that the box doesn't allow you to see all the text. Unless you are a very good typist, it will be easier to type your Headline in a word processor or text editor. Then when you are confident that there are no misspellings, you can erase all the text in the box. Copy and paste your new headline in. Then save your work.

There's one more thing you should consider when editing the Basic Information at the top of your profile. LinkedIn will display your name, headline, **and** location on its search results page. You can adjust what information LinkedIn shows in the results using the **See what other users in your industry are using** link located under the **Show examples** link highlighted below.

Once you change your headline, be sure to click **Save** before leaving the page.

Choose Skills

You may include up to 50 skills in your profile. You'll want to choose skills that your connections can verify. In addition, stick with skills that enhance your value. For example, John Wordsmith may be a great cook, but those skills have nothing to do with his writing career.

---

4 Top Headline Tips

When someone searches for you on LinkedIn, they see three things: Your name, your LinkedIn headline, and your location. Expect the decision to learn more about you to rest on the impression these three elements make on the hiring manager or recruiter.

1 **Compare your professional headline to a newspaper or magazine headline.** A headline should always hint at what will follow. The reader should have an idea of what your profile will include.

2 **Be specific.** It results in a much better headline. If you are vague, you won't grab the attention you desire.

3 **Great headlines attract attention.** The more people who view your profile, the better your chances of connecting with the right person who can lead you to your dream job.

4 **Identify your skills set.** Your headline needs to quickly identify you as a certain type of person – i.e., manager or executive, or someone who specializes in a certain field or industry.

A well-written headline will help you to structure the rest of the information you include in your LinkedIn profile. If the information doesn't support the headline, consider whether it should be included at all. Remember, focus is important.

NOTE: LinkedIn's default headline setting is your job title and the company you currently work. If you don't change it, this is what LinkedIn will show on your profile.

As you start typing in a skill, LinkedIn will recommend related skills. You'll want to use these because they are already in LinkedIn's keyword bank. That means people are using them to find others.

Just because you can add up to 50 skills doesn't mean you have to. Keep focused on your goals. You don't want to come across as a "Jack of all trades and master of none" type. Emphasize the skills you do best.

Once you list your skills, your connections can endorse them. Your connections can also endorse you for skills you haven't thought of on your own.

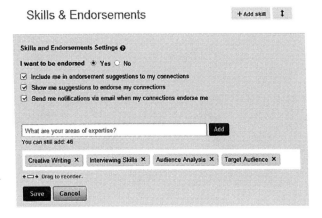

You have the choice of showing your endorsements or keeping them hidden.

LinkedIn will display up to 99 connections who have endorsed you for a skill.

Today we live in a click-happy society and it is very easy to endorse a connection for skills when we view a profile. Because of this, endorsements are not as valuable as recommendations.

Describe Your Current and Past Positions

LinkedIn's advice to choose two or three things to emphasize about each position is a solid strategy. Most people try to stuff too much information into this section.

Emphasize those things that made the greatest difference in that position. For example, as a resume writer, what would you want to know about me before you hired me?

Likewise, with your previous position, emphasize as much as possible those aspects of the job that enhance the perception that you have the skills to write the kind of resume that gets results.

You are laying a foundation for the types of references you will look for at a later stage. You can seek out prior clients or colleagues that you've helped or who will validate your expertise and value.

Rather than trying to build yourself up, think about what others will say about you, and use that as a starting point. You can always come back and revise based upon the references you actually receive.

You'll be given a new list to check off after you share your profile (or opt out). LinkedIn will:

- Check to see if you have any additional education to add to your profile.

- Ask you if you know any other languages.

- Ask you to add a summary of who you are and your objectives.

Experience

+ Add position

Staff Writer

Fortune 500 Company

January 2006 – Present (9 years 3 months) | Iron City, NC

Manage the creative writing lifecycle process for a Fortune 500 publishing company.
~ Conceptualize ideas and create engaging stories with an angel.
~ Interview with internal and external Subject Matter Experts.
~ Manage the editorial calendars and performance of 11 internal and freelance staff members.

Additional Enhancements

- Give you the opportunity to upload projects you've worked on.

- Give you the chance to list coursework you have taken.

- Allow you to add publications, test scores, patents, and volunteering and other causes.

Write Your Summary

When you come to the summary question, you'll want to expand the size of the text box. Look for the little arrow at the lower right corner. Mouse over the corner until it turns into an arrow. Then you can click and drag the window to expand the white text area. Doing this will make it easier to edit your content.

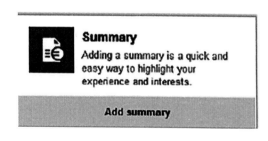

Summary

Adding a summary is a quick and easy way to highlight your experience and interests.

Add summary

After the headline, the most-often read section of your

profile is the Summary. LinkedIn allows up to 2,000 characters in your summary. Use them wisely—and use all of them!

One format that is very effective is the Who/What/Goals structure. You begin with **Who** you are, **What** you have to offer (what is unique about you or your experience?), and what **Goals** you have for your career or for being on LinkedIn.

You may repeat this pattern numerous times throughout the summary by dedicating one sentence to Who, one to Why, and one to Goals.

Another effective formula is to shift to a how focus after you have written your Who/What/Goals opening. This is especially effective for consultants and service providers.

You might find it difficult to use all 2,000 characters at first. Don't worry about it. It is better to write a tight summary than to ramble on to meet a character quota. In the example shown in the next screenshot, there are just over 1,000 characters.

Remember, you can always revisit your summary and expand it. Just hover the cursor on **Profile**, then click **Edit Profile** in the navigation bar. You'll be able to click the blue pencil next to **Summary** on the page that loads.

Summary

* CORPORATE COMMUNICATIONS & MEDIA RELATIONS: Trusted advisor, credentialed coach, and savvy strategist of corporate communications and media relations. Audience seer and dragon slayer of employee trances caused by boring, corporate-lingo. Fan of messaging that use purposeful talking points, presentations, content marketing and brand journalism pieces that zero in on what people care about. Develop press releases, media inquiries, articles, position papers, opinion pieces, etc. Write to express, not impress!

* COMMUNITY OUTREACH & EMPLOYEE ENGAGEMENT: Leader of measured engagement and recognition experiences that revitalize soul in the workplace and mirror a company's values to gain trust. Natural behaviorist and neutral diplomat in pinpointing breakdowns and rebuilding bridges that form a culture everyone wants: inspiring, happy, transparent, high-performing, and yes, profitable.

* HR, TALENT DEVELOPMENT, & RECRUITMENT: Restoring "human" back into "resources" for companies that realize you can still mitigate risk while nurturing a culture of humanity and autonomy. Sharp trouble-shooter with a sixth sense in identifying where a company is having trouble with the curve. Credentialed coach with rich experience in talent planning and development, change management, post-merger integration, recruitment, and behavior/career assessment and interpretation. Let people master what they do best!

Additional Summary Writing Tips

The first two or three sentences need to instantly get your prospects interested in your profile—or, even better, get them excited about reading the rest of your profile. Your LinkedIn summary can set you apart from other job seekers on LinkedIn by demonstrating that you understand what employers want and what you have to offer that meets that need.

To do this you must address specific questions every prospect is asking:

- **How** will you add more value to the company, or solve problems better than other job candidates?

- **How** will your next employer benefit by hiring you? Quantify the value in terms of numbers, money, and/or percentages. Use specific numbers and facts to build credibility.

- **What** experience can you offer that will provide value to your next employer?

- **What** additional skills do you have that set you apart from other candidates with a similar background?

Write naturally and conversationally. In contrast to your resume, it is okay to use pronouns in your summary if you wish. Regardless, whether you use or exclude pronouns, speak in the first person, not third person.

For example, "I did such and such. Write as if you're speaking to an individual reader. Make it personal. Be sure to emphasize outcomes—as well as what makes you uniquely qualified to do the job you do. Try to find a common thread that connects all of your work experiences.

Then, once you have a theme, use storytelling principles to write your summary as a narrative. Have a beginning, a middle, and an end.

Your summary can be anywhere from a few sentences up to a few paragraphs. Nevertheless, don't waste any words. Make the most dramatic, powerful, attention-getting statement possible.

Don't use more words than necessary, and avoid flowery language.

The point of the first sentence is to get the prospect to read the second sentence ... the next one ... and the next.

Summary

In the competitive world of professional cycling, the competition isn't limited to the athletes in the peloton. The teams themselves are also competing -- for sponsors, fan attention, media coverage, race invitations, and social media status.

In my role as Team CycleProSports' National PR Rep, I grew the team's social media presence by more that 25% in one year. This included facilitating more than 12 million fan engagement opportunities online. I also secured stories in all the major cycling industry trade outlets and consumer-facing publications, including "ProCycling Monthly," "Biking News," "Two-Wheeler," and "ProVelo."

I also handled crisis communications for the team, including the controversial finish of the Tour of Africa, and media relations after the bankruptcy of one of the team's major sponsors. The team's owner recognized this effort with a handwritten letter praising my "responsiveness, tact, and diplomacy in effectively handling what could have been a disaster" for the team.

I am currently pursuing offers for the 2012-13 season. (As an independent contractor, I work exclusively with one team each year on a contract basis.) If your professional cycling team is looking to increase your team's profile in the media, with fans, and online, I would love to talk with you about what I can do for you. Click on the website link in my profile for full contact information.

I am aslo interested in connecting with you on LinkedIn if you are a journalist or blogger who covers professional cycling, a race organization official or sponsor, or UCI- or USA Cycling-accredited official.

Be conversational and informal in your tone. Use contractions ("you're" instead of "you are"). Every word counts! Pay attention to grammar and spelling. Make sure there are no mistakes in your profile. Re-read and edit it. Have a colleague, friend, or spouse read it. Copy and paste it into a word processing program and run a spell-check on it.

You can also use asterisks, dashes, hyphens, and other keyboard characters to format the summary and make it easier to read. Notice these elements in the Summary below written by Fred Jobseeker.

Notice the format:

- In the opening paragraph, draw attention to issues, challenges, or problems faced by your prospective employer.

- In the second and third paragraphs, demonstrate the value you offer to employers by quantifying the accomplishments in your current position (ideally related to the problems outlined in the first paragraph).

- In the fourth paragraph, talk about why you might be open to inquiries (if you are a passive candidate). If you are unemployed, you might state the reason your most recent position ended (if the company closed, for example) or that you are available immediately.

- Give the reader information on how to contact you. (Note: LinkedIn's Terms of Service prohibit you from providing your email address directly in this section. Instead, direct them to connect with you on LinkedIn or use one of your links to provide a method for direct contact.) You can also use the Personal Information section to provide a phone number.

Using these strategies, you can develop a LinkedIn headline and summary that will lead to job opportunities, contacts from prospective employers and recruiters, and increased visibility online for your job search.

Upload Work Samples or Projects

There are two ways to highlight different projects. You can upload a file or insert a link (video, images, other documents, PowerPoint, Slide Share, etc.) as part of the Summary section or under each job of the Experience section.

This is a way to highlight visually and/or audibly some of the things you have done.

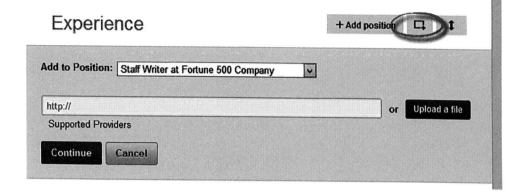

The second way is to fill out the Projects section. If you choose to link to a project, LinkedIn will ask you to create a title for your project. Choose a title that uses keywords for your industry, if possible.

At the same time, don't sacrifice clarity on what the work sample will show visitors to your profile. Keep the title on target for the content.

LinkedIn also allows you to add your connections who were involved in the project with you and highlight your whole team.

The Projects section is the perfect place to share examples of your work. If you don't have any projects, you may skip this section, yet it really is worth creating a special work sample or project just for LinkedIn if you are in certain careers.

For example, a writer wouldn't want to miss this opportunity to share examples of articles that have landed notoriety for a client. (Of course, unless it was published, she or he would change identifiable information).

Add Courses

Now, you have the opportunity to highlight courses that demonstrate you have domain knowledge of your field.

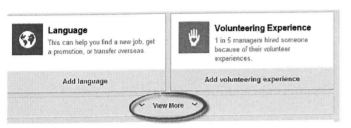

These courses don't have to be from one of the formal educational institutions you've attended.

While this is optional, it's a good place to include training you've taken that isn't necessarily issued by an educational institution. Perhaps you completed employer-based training. You may add independent coursework or classes taken through an external association.

To add more than one course, you'll have to visit **Edit Profile** from the navigation bar. Scroll down to click View More to expand the list.

Once the list expands, find the Courses category and click **Add Courses**. LinkedIn will open the screenshot shown, which will allow you to add courses.

Once you add your courses, it will look like the screenshot below. It will also give you the message that as far as it's concerned, you are done, yet you still want to keep on improving your profile. If you click **Continue**, LinkedIn will take you through any steps you overlooked before.

You'll also get a second chance to enter more courses if you have more to add. You may also notice an additional category from the **View More** pull-down menu—Publications.

It is highly recommended that professionals publish at least one book connected with their expertise. Today, there are many venues

Publications

Publications are a great way to show off your professional accomplishments.

Add publications

of self-publishing and eBook opportunities that provide professionals a voice in his or her chosen profession.

To add publications, follow the same process that you did when adding your courses.

If you have not published an eBook through one of the main publishers such as Kindle or CreateSpace, you can provide the URL to the page from which potential clients or customers can download the eBook. If you don't have a URL for your publication, click **No URL for this publication**.

At this point, you have completed the meat of your profile. If you see other categories that you would like to add notable accomplishments, by all means, do so. As you continue to use LinkedIn, it will ask you to add more positions.

If you don't have additional positions to add, there is one strategy you might be able to use. For example, if you have worked for the same business for 10 years, yet changed positions within the business during that time, consider breaking up your experience to reflect this. It demonstrates progression.

Another area that LinkedIn will consider incomplete is your connections. You will have to continue looking for connections until you have met the minimum quota.

Complete your LinkedIn Profile. It is important!

Use this Checklist to Make Sure You Haven't Missed Anything

Basic Profile Activities

- ✓ Profile photo
- ✓ Your current industry
- ✓ A current position with description
- ✓ Two or more positions
- ✓ Education
- ✓ At least five skills
- ✓ At least 50 connections
- ✓ A summary

The level you are at on the right side of your Profile page reflects how close you are to finishing these items.

Most Important

- **Proofread** your profile carefully. Check grammar and spelling!

- **Update your profile regularly.** Not only will your connections be notified when you update information on your profile (bringing your profile additional visibility), but you'll also be confident that someone searching for you will have access to the most current information.

Job Search Specific Activities

- Customize your LinkedIn profile URL (www. linkedin.com/in/yourname)

- If you're including a link to your website or blog, customize the text link (rename it so it doesn't just say "Personal Website" or "Company Website").

- Include your contact information. LinkedIn allows you to add your phone number (designated as home, work, or mobile), Instant Messenger contact information (AIM, Skype, Windows Live Messenger, Yahoo Messenger, ICQ, or GTalk), and multiple email addresses (in addition to your primary/sign-in email).

- Add languages that you speak.

- Fill in key projects you've worked on (this is a separate section within the profile).

- Add a list of courses you've taken. (This helps with keyword searches.)

- In the Settings, change the Select what others see when you've viewed their profile to your name and headline (recommended).

Take Action!

Focus on completing your profile so you can achieve All-Star profile strength. You only asked for connections one week ago, so you have to be patient. Not everyone signs in to LinkedIn every day. It will take some time to meet LinkedIn's connection quota.

Keyword Basics

Keywords play an important part in helping people you may not know, find you—this is particularly true for job seekers who are hoping for contacts from prospective employers and recruiters.

LinkedIn headlines are searchable fields using the "People Search" function when someone is looking for particular skills, interests, qualifications, or credentials. They help others find you!

What are keywords

Keywords are words and phrases related to your work. They help a prospective employer find you when they need someone with those skills.

Where can you find keywords?

Brainstorm them. Write down a list of words that relate to you, your work, industry, and accomplishments. Try to come up with as big of a list as you can; you will narrow it down later.

You can also find keywords in job postings or job descriptions. Check out online job boards for positions. Don't worry about where the job is located; just find positions that are similar to the one you're seeking and write down the keywords.

You can also find broad job descriptions—with plenty of keywords—in the U.S. Department of Labor's free Occupational Outlook Handbook (**http://www.bls.gov/ooh/**).

Another great research tool is Google's AdWords Keywords Tool, which is found at: **https://adwords.google.com/select/keywordToolExternal**. You can use keywords you identified through your earlier research. Google AdWords suggests related keywords and tells you how popular those keywords are in current Google search results.

How do you choose keywords?

You need to pick the Top 10 keywords that you will use in your LinkedIn headline and profile. The keywords that you select for your profile must fit two criteria:

- They must speak to your "onlyness"—what you want to be known for.

- And they must align with what employers value— what they want.

Focusing on these areas enables you to get the most out of your online efforts while differentiating you from other job candidates with the same job title. You need to express clearly: "I am this." Someone who is reading your LinkedIn profile should be able to recognize you in it. If what you wrote could apply to anyone with your job description, revise what you've written.

PRIVACY

Y ou have the option of restricting certain parts of your LinkedIn profile so the public cannot view them. In the past, LinkedIn users had more control over which LinkedIn members could view their profile. Now, the restrictions are more limited, and when you change your privacy settings, LinkedIn is referring to your privacy across Google or other major search engines.

When you are in active job search mode, you want your privacy settings to be as open as possible to encourage more profile views and connections. Google indexes LinkedIn profiles, so having an open profile can gain additional visibility through Google. A public profile on LinkedIn presents your professional skills to those who want to know more about you.

However, if you are currently employed, you want to consider your privacy settings more closely. You might want to limit viewing of your profile to direct connections or connections within your network.

Changing Your Privacy Settings

Change your public profile display by hovering your cursor over Profile from the navigation toolbar. Choose **Edit Profile** from the drop-down menu. Then scroll down until you see the website address that contains your name. Click the **edit link** to the right of the website address.

The information that matters to you is on the right. You'll see **Customize Your Public Profile.**

Notice the options you can deselect from public view. To obtain maximum exposure as an unemployed job seeker, make your public profile visible to everyone. When your job search is complete, you can adjust your privacy

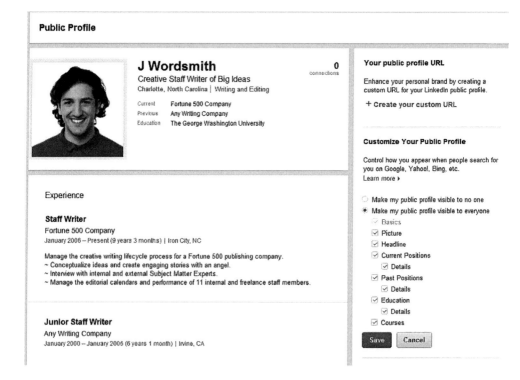

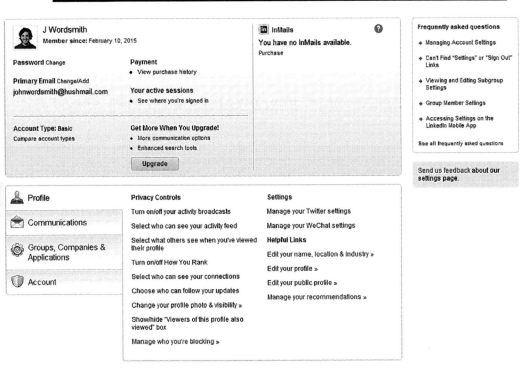

If you click **Turn on/off your activity broadcasts**, a window pops up. If you uncheck **Let people know...,** you won't automatically send out notices that you are looking for a new job, if you don't want your current employer to know about it.

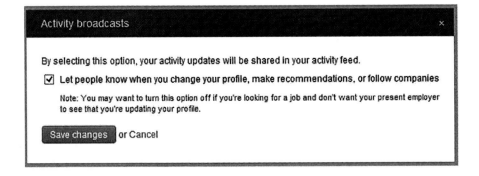

Clicking **Select who can see your activity feed** changes

whether or not a visitor to your profile sees all of your recent activities below the advertising area. The default setting is **Your connections**, but you may want to choose **Only You** while you are building your profile because it will generate messages each time you update the profile, and this offers no value to your visitors.

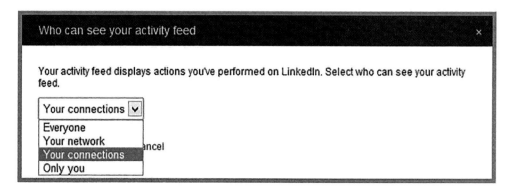

Once your profile is fully developed, you should revisit this option. If you are in an active job search, **Everyone** may be a wise choice if your network hasn't resulted in a job offer.

Another setting you can change is what others see when you've viewed their profile. You can be totally anonymous if you wish; however, there is a caveat to this. It means that you cannot see who has viewed your profile. If you are in an active job search, use LinkedIn's recommended setting.

LinkedIn gives you two options when it comes to whether people can see your connections: **Your connections** or **Only you**. Again, there is another caveat. If you share a connection with the person you do not want to see your profile, they will be able to see it through that shared connection. In an active job search, you want connections to be visible to your connections.

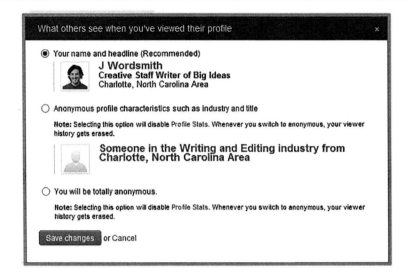

Change your profile photo and visibility is self-explanatory. Making your picture visible to everyone is the best practice if you are actively looking for work.

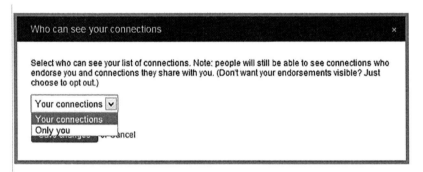

Choosing Communications Preferences

These settings are worth paying attention to because they control how accessible you are to the LinkedIn community.

The **Member Communications** section allows you to select the type of messages you're willing to

receive. You'll see that LinkedIn defaults to **Introductions and InMail only**, and for **Opportunities**, LinkedIn checks all opportunities. You can personalize what types of opportunities that interest you.

Take the time to write a brief message to potential contacts if you are seeing job inquiries or consulting offers. This helps to establish reasonable expectations.

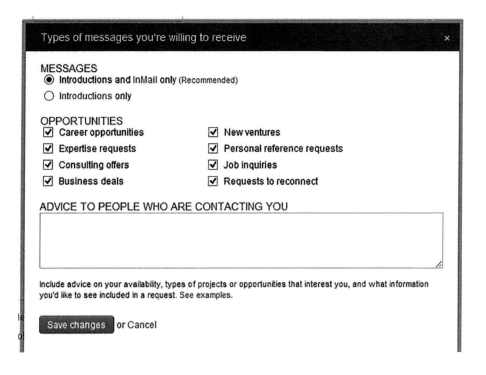

Showing who viewed your profile during a job search could be useful, but isn't essential. If you prefer to hide this information, click **Show/hide "Viewers of this profile also viewed"** box. This will remove the check.

Viewers of this profile also viewed...   ×

☐ Display "Viewers of this profile also viewed" box on my Profile page

Save changes or Cancel

If you have a Twitter account, you may choose to add it. If so, make sure you use that Twitter account professionally, not to share your personal life.

InMails and Introductions

Avoid missing important InMails and Introductions by customizing how you want to be notified when you receive them.

Emails and Notifications
Set the frequency of emails
Set push notification settings

Member Communications
Select the types of messages you're willing to receive
Select who can send you invitations

LinkedIn Communications
Turn on/off invitations to participate in research

Turn on/off partner InMail

Review each option. If you are in an active job search, select **Individual Email** for each category. This may mean a lot of emails, yet it's the best way to stay in the loop and respond promptly to any opportunities that arise.

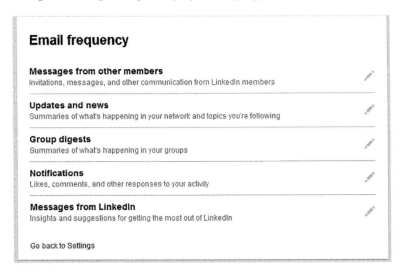

Email frequency

Messages from other members
Invitations, messages, and other communication from LinkedIn members

Updates and news
Summaries of what's happening in your network and topics you're following

Group digests
Summaries of what's happening in your groups

Notifications
Likes, comments, and other responses to your activity

Messages from LinkedIn
Insights and suggestions for getting the most out of LinkedIn

Go back to Settings

If you aren't actively looking for work, then you will find the weekly digest option reduces the flow of messages into your inbox. We recommend that you keep the following settings at **Individual Email:**

- InMails, Introductions, and Open-Link

- Invitations

- Questions from your connections

- Replies/Messages from connections

STRATEGY TIP

While you are building your LinkedIn presence, it's a good idea to make your profile visible to no one. When you have uploaded a picture, written a good headline and summary, and filled out current and past positions, then make your profile visible to everyone.

Personalize the remaining settings, recognizing that frequency is good for where you are actively interested in opportunities.

Invitation Filtering

When you are in a job search, you should definitely allow contacts outside of your network to get in touch with you, so don't change LinkedIn's default **Anyone on LinkedIn** setting under **Who can send you invitations.** In an active job search, you should initially accept all invites to increase the chances of growing your connections. Group digests will matter when you've joined a group. For now, you can leave this setting at its default.

Emails and Notifications
Set the frequency of emails
Set push notification settings

Member Communications
Select the types of messages you're willing to receive
Select who can send you invitations

LinkedIn Communications
Turn on/off invitations to participate in research
Turn on/off partner InMail

LinkedIn has its own set of communications. Review

these and set them to match your personal preferences.

Whether you have these turned on or off will not affect your job search.

EXPERTS' TIPS

If you do not have any connections in common with your target contact, it's a good idea to identify if that person participates in any online forums you can also join. For example, does he or she use Twitter regularly? You can connect there, share tweets, and engage enough so the person know about you. Or, find out if the contact is active in a LinkedIn group. Once you identify what network the person uses and prefers, you can engage and create a "warm lead."[1]

Without sounding like a broken record, it's very simple to reach out people you've never met. Be curious about them. Before you reach out, consider "what's in it for them?" If you can't think of a reason this person would want connect with you, dig deeper and find one.[2]

---

1   Miriam Salpeter, job search and social networking coach, Keppie Careers (www.keppie-careers.com). Author of 100 Conversations for Career Success (with Laura Labovich) and Social Networking for Career Success.

2   Laura M. Labovich of ASPIRE! EMPOWER! Career Strategy Group (http://aspire-empower.com/). Author of Two Weeks to Job Search Discovery!

What to Do with Your LinkedIn Profile

You've built your LinkedIn profile. Are you wondering, **"Now what?"** (There's an excellent book that answers that question: *I'm On LinkedIn, Now What???* By Jason Alba. You should check it out.)

Status updates and announcements from your personal network will post to your Home page. At a glance, the **Updates** section will keep you on top of what's going on with your connections. To make the most of your interactions on LinkedIn, this is an area to check on regularly.

You can respond to individual updates right on your Home page—you can retweet, favorite, reply, like, comment, or share updates.

To broadcast your own updates, input messages into the status box and click **Share** to send messages to your network.

What kind of information should you post in your updates?

- Current and upcoming trends in your industry
- Insights from projects you're working on
- Events and seminars you've attended
- Training courses you're taking

- Links to articles/blogs within the industry

- Inspirational quotes

The content that you share doesn't necessarily have to be yours 100 percent of the time. It's a great idea to share links to content from others in your industry, along with your thoughts on how this work will affect your industry.

You can post several updates each day or one every few days. At a minimum, you should post a new update at least once a week.

MAKING CONNECTIONS

LinkedIn organizes your relationships with other LinkedIn members using a tier-based system. This means that people who acknowledge that they know you by connecting with you become bridges to additional connections.

There are three levels outside of your network. Level 1 is assigned to anyone you are directly connected with. Level 2 is assigned to anyone your direct connections are linked to. Level 3 is assigned to people who are connected with Level 2 connections and not Level 1 connections. The greater the connection level number, the further away from you a connection is.

You can view all 1st level connections and the number of people you share as connections and their identities.

When you perform a search, you'll also see if any of the search results show people connected to you, and at what level.

Knowing the tier a connection falls on determines your strategy as you work toward growing your network.

Import Contacts

Looking for people one by one wastes valuable time that should be spent on a job search. You could use LinkedIn's search box to locate possible connections, but LinkedIn provides several ways to streamline this process. Once you master these techniques, you'll be ready to move on to connecting with people you don't already know.

For those who use Hotmail, Gmail, AOL, or Yahoo, learning how to import contacts from other email management systems will be a major timesaver. Let's begin by reviewing how to import contacts from one of the main web-based email services.

How to Import Contacts from Hotmail, Gmail, AOL, and Yahoo

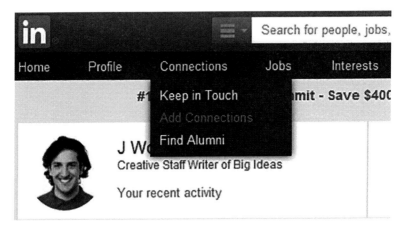

To import contacts, choose the **Network** menu from the navigation bar. Select **Add Connections** from the dropdown menu. There, you'll find several choices for importing data.

Whether you decide to have LinkedIn sync with an online email, manually enter contacts yourself, or import from a desktop email, this networking tool makes it simple.

Notice you can also click your alma mater and find connections that way. If you've entered the date of graduation, LinkedIn will prioritize the connections by suggesting graduates in a similar time range.

See Who You Already Know on LinkedIn   Manage imported contacts ▸

Gmail	Outlook	Yahoo! Mail	Hotmail	AOL.	Any Email

Get started by adding your email address.

Your email

[   ]

[Continue]

🔒 We'll import your address book to suggest connections and help you manage your contacts.
Learn More

If you have clicked **Add Connections**, you'll see seven different options to choose from.

Note: The Outlook button only works for businesses that allow their employees to log in to their Outlook account remotely. It won't work if you are managing email using Outlook on your home-based business computer.

The **Any Email** option works with many of the common email services around the world.

You will have the opportunity to approve and decline the additions of contacts as LinkedIn searches the designated online email box.

How to Import Contacts from Other Email Applications

You can also upload .CSV, .TXT., and .VCF files. Most desktop contact management applications—like Outlook—let you export addresses to one of the file types mentioned.

You'll find the upload link on the **Any Email** page, below the boxes for entering your email and email password. Scroll down the page if necessary until you see **More ways to connect.** Click **Upload contacts file.** This will give you access to the **Browse** and **Upload File** links.

See Who You Already Know on LinkedIn   Manage imported contacts ▸

Get started by adding your email address.

Your email

Continue

🔒 We'll import your address book to suggest connections and help you manage your contacts.
Learn More

Of course, you need to create a .CSV, .TXT or .VCF file to upload. I've included the instructions for Outlook below. If you have a different email application, click **Learn More.**

LinkedIn's exact instructions as to how you import your address book are as follows:

1 Move your cursor over **Connections** at the top of your homepage and select **Add Connections.**

2 Click the button for the email provider you use. If you don't see your provider, click **Any Email.**

3 Enter your information if not pre-populated.

4 Click **Continue.**

5 Contacts who are already on LinkedIn will be shown. Click the **Skip this step** link if you don't want to invite anyone OR click **Add Connections** to send invitations.

6 Contacts who are not yet on LinkedIn will be displayed next. Click the **Skip this step** link if you don't want to invite anyone OR click **Add to Network** to invite them to join.

*Note: LinkedIn states that it will automatically select all contacts on the displayed list to be invited. If you don't want to send invitations to everyone on the list, be sure to uncheck the **Select All** box and individually check the boxes next to contacts you want to invite.*

Once your address file is uploaded, you'll be sent to a page that allows you to choose who you want to ask to connect with you. You don't have to invite every- one in your list. Deselect anyone who isn't some one you have actually worked with or know well.

STRATEGY TIP

If you work in MS Excel on your computer, you may have hundreds of *.CSV files.

If you do any computer work, you will have thousands of *.TXT files. In fact, owning a PC guarantees you will have a hard time locating your file. TXT is a common computer extension for PC software instructions.

Email addresses for people who have pro- vided product sup- port, for example, aren't likely to remem- ber you, so an invita- tion from you might appear to be spam.

Also, does it need to be said? Deselect anyone who you know won't appreciate an invita- tion from you.

It is also worth checking on the number of contacts the person has as you are making your decisions. A person who has been a member for a year and only has four connections isn't going to be a valuable contact. These individuals have started on a LinkedIn profile, yet aren't willing to put any time into developing their network.

At the same time, if you see someone who can benefit you with a reference, invite them regardless of the num- ber of connections they currently have.

Note: Importing information does not automatically connect you with people. After importing, each person will receive an invite from you to join your network on LinkedIn. Once the recipient accepts the invitation, then you are connected.

When building your presence on LinkedIn, you need to start somewhere. By requesting connections on LinkedIn with people you already know—either in real life or through email correspondence—you can build your network, which will help you realize your job search objectives.

How to Search for Connections

To the right of the navigation toolbar is a search box. Click the **Advanced** hyperlink to the right of the box.

The **Find People** option will identify likely contacts as you scroll down. Your existing contacts are at the top of the list.

You can also use the **Advanced People Search** function to identify contacts. You can use keywords to search by specific job titles. You can use the variables "Not" and "Or" to define exactly what you are looking for. If you are looking for a local person, or someone who shares certain interests, you can specify this in your search terms.

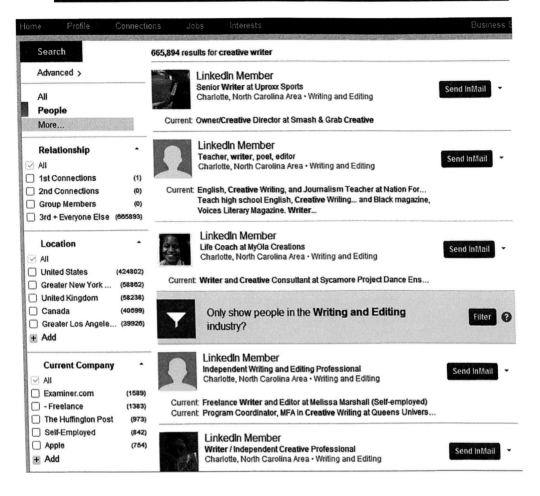

You'll notice that you can search by companies you've worked for, which will help you find former colleagues. You can search by school to find former classmates.

You can also narrow your search by industry. This can be useful if you are searching specifically to grow your network within in your niche.

You can search for only 2nd connections. As you will learn later, 2nd connections can prove important for gaining exposure within a specific job market. Many times, people connected to your connections can be affiliated with markets that interest you. We'll talk about introductions later.

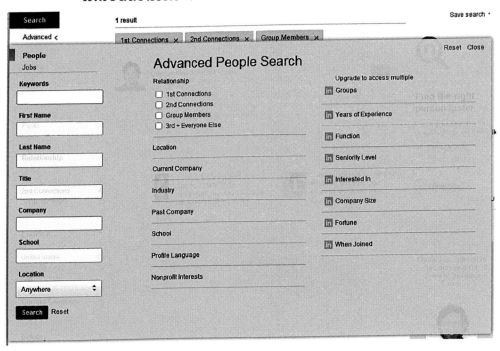

Ways to Use Search Results

Now that you've found people, how do you connect with them? What do you do with the search results LinkedIn delivers? Regardless of whether you're directly on a profile page or reviewing a listing of profiles from a search, these are the options usually offered to initiate contact:

- Send InMail
- Get introduced
- Find references
- Share Profile

POWER TIP

Once you have your search criteria set, you can the results. These will then be placed under the "saved Searches" tab. You can re-run this search as often as you wish. With a basic account, you can save up to three searches. You can even have LinkedIn run te search on a regular weekly or monthly basis and email you the results.

If you mouse over the arrow beside the **Connect** button or **Send InMail** button, you'll see what your options are. When the person is a 2nd tier connection, LinkedIn allows you to request a connection directly, yet it is much better to select the **Get Introduced** option. This way someone who knows you is recommending the connection.

If you have had a good business relationship with the person who is introducing you, he or she is quite likely to make the introduction gladly.

When the person you are interested in connecting with is a 3rd tier connection, you'll see the option to **Send InMail.** If you have a free membership, you'll have to upgrade to a Premium account in order to send an InMail. You can receive InMails with a free account, but not send them. It's better to use the **Get Introduced** option for this type of request, as well.

POWER TIP

Use LinkedIn to research your current colleagues. You can use it to learn more about your fellow employees.

Share Profile is available to you when you want to recommend a connection between two of your connections. Be very careful about recommending connections between people you don't know verse those you do know. Make it very clear that you don't know the person you are recommending.

For example, if a writer was too busy to take on a project for a contact, a person might recommend that the person check out the qualifications of another writer with an impressive profile.

One of the cool things about the **Find references** option

is the way it searches your network for anyone who may have worked with someone you are researching.

If you find you can't get the "We're sorry window" to go away, just click the search button again.

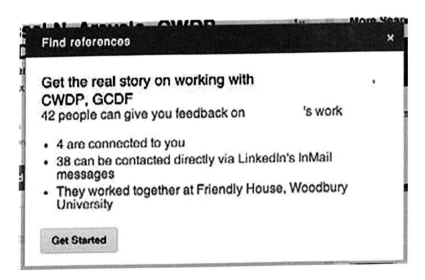

Find references   ✕

Get the real story on working with CWDP, GCDF
42 people can give you feedback on   's work

- 4 are connected to you
- 38 can be contacted directly via LinkedIn's InMail messages
- They worked together at Friendly House, Woodbury University

Get Started

More about Introductions, InMails, and Invites

When you are viewing someone's profile page, you'll see your connection options right below the brief overview LinkedIn provides. When working from a search result page, you'll see the contact methods to display on the right side beside each member. Below are additional strategies you can use to reach potential connections.

How to Use Introductions to Make Connections

Facebook has "likes," Twitter has "re-tweets," and LinkedIn has "introductions." Designed around the principle of referrals, introductions are one of the best ways to meet new people on LinkedIn. Introductions are highly regarded because they deliver built-in trust.

The basic idea is that if you don't know a person in which you would like to connect with, find someone within your personal network who can introduce you to this person. When a potential contact sees you linked to someone s/he knows, it implies you are a credible person.

A search for writers within a specific area reflects a 2nd tier connection. By clicking on **shared connection**, you can identify the common link. This is a good opportunity to reconnect with a tier 1 connection and ask for an introduction to that connection. To do this, you will click **Get Introduced.**

This takes you to a new window where you can see yourself on the right, as well as the name(s) of the person(s) connecting you in the middle, and the person you would like to be introduced to on the left.

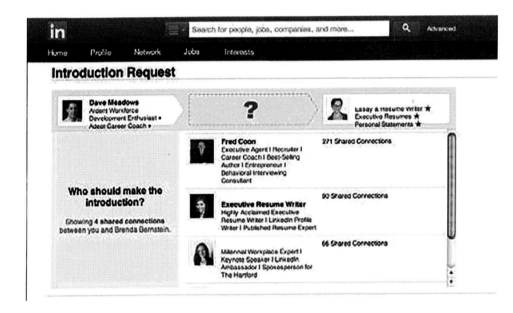

Select the person you want to make the introduction. You will go to a new page.

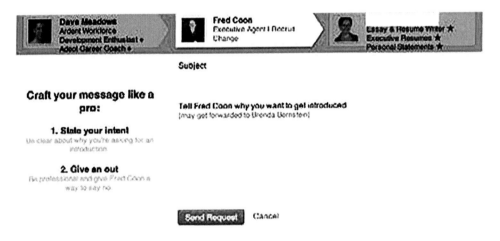

Take the time to write a good explanation for why you

are reaching out and why you would like to make the connection.

For example, for the above Introduction Request—

Hi <name of person>,

I hope your work as a PA has been going well. I have been focusing on developing my resume writing skills over the last six months, including adding LinkedIn training to my services.

As I was preparing a lesson, I found <name of person I want to connect to> in a search for writer and found that we share a connection. Would you mind introducing me to <first name>? At times, I'm not able to take a client, and I'd like to expand my network with like professionals.

Thanks, Fred

Be sure you enter a subject or LinkedIn will send you an error message.

How to Use Connect

In general, when a person is at the 3rd level in your network, it isn't worth cold connecting. It's too easy to be labeled a spammer. Nevertheless, if you share a common career path, it can be a good strategy to grow your connections through carefully written invitations.

For example, perhaps the 3rd tier professional is someone you have spoken with on teleseminars and communicated with via email. This individual may not use the same email on LinkedIn as you have in your contact list. This would be an ideal person to reach out to and ask for a connection. In this scenario, it makes the most sense to click **Connect.** In the window that pops up, you'll have several choices.

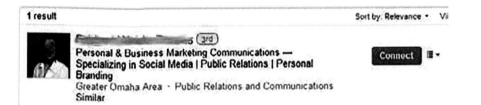

In this case, you might choose **Other** because you are a Colleague. Also, **Other** fits more appropriately at this stage because this connection has been more of a mentor, which of course, is not listed.

When you click **Other,** LinkedIn may ask for an email address. This is to ensure you aren't just spamming the individual. This may also happen when you click other options.

The key thing to remember is this: don't assume the person will remember you. Trigger his or her memory with something that will create a connection. Personalize the message! It will pay off in more connections.

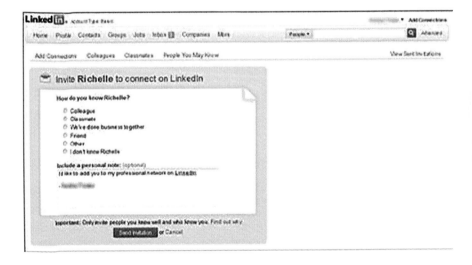

How to Use InMails to Connect

InMails are a paid feature of LinkedIn that enable you to send emails directly to a person's LinkedIn mailbox, regardless of whether or not they're in your network. You have to have a premium (paid) account in order to send InMails. Depending on your premium account level, you'll receive a credit for a certain number of InMails per month. If you send an InMail but don't get a response from a recipient after seven days, you will still be credited for the InMail; however, you won't lose a credit for InMails that do receive a response.

STRATEGY TIP

Indicate that you like an update. Leave a comment. These are discrete ways to gain the attention of people in your network who may be hard to connect with otherwise.

Although InMails require payment—and you're running a risk by contacting someone who doesn't know you personally—they can be an effective way to connect with someone who you don't know directly.

Using InMails to meet a fellow LinkedIn members shows you're serious about your job search and willing to invest to make the right connections.

Messages

When you are already connected to the person, you will have different options than when they are not part of your network. The primary option that changes is the ability to message on LinkedIn. This service is free.

How Make Inroads with Invites

Technically, you can send a LinkedIn invite at any time. You just click **Connect** beside any search result. With the **Invite** feature, you don't need a connection or a paid account. However, use discretion with this method. Recipients can respond by stating they don't know you and preventing you from ever sending another invitation. If you receive too many "I don't know this person" responses, LinkedIn may restrict you from sending invitations altogether.

Always change the default text when inviting someone to connect. The default phrase is, "I'd like to add you to my professional network on LinkedIn." Change this to highlight a personal connection in- stead. For example, "I enjoyed meeting you at the workshop yesterday. I'd like to connect with you on LinkedIn to keep in touch."

Take Action!

Follow the steps given to import your existing contacts into LinkedIn. Then, use the **Advanced Search** function discussed in this chapter to see who else pops up that you can connect with. Next, run and save a **Saved People Search.**

Post a status update after you have completed these instructions.

Plan on posting at least one status update each day this week by doing something from this or one of the previous lessons that you want to share.

Checklist for Maximizing Your Connections

Click on the Add Connections link in the upper right hand corner of any page to:

✓ List your email addresses to find people you've sent messages to who may be on LinkedIn.

✓ Import contacts from Outlook, Apple Mail, and other email clients that keep lists of clients.

✓ Send invitations to individuals by entering their email addresses from the Add Connections page.

✓ Search for past or present colleagues under the Colleagues tab, based on current and past positions you've listed in your profile information.

✓ Search for past or present classmates under the Alumni tab, based on current and past educational institutions you've listed in your profile.

✓ Review the People You May Know tab for direct and indirect contacts based on your profile information, interests, and communications.

You can also access the Skills tab (under the More button in the navigation toolbar) to identify professionals in your field(s) of interest who have expertise related to your work. This is a good way to establish connections with other experts in your line of work.

REFERENCES FOR BUILDING CREDIBILITY

References are a powerful tool in a job search. Recommendations on LinkedIn serve much the same purpose. A recommendation is "social proof" from a third party that you're a skilled professional.

According to LinkedIn, "Users with recommendations in their profiles are three times more likely to receive relevant offers and inquiries through searches on LinkedIn."

In fact, if LinkedIn is still telling you that your profile isn't quite complete, this could be the primary reason.

LinkedIn recommendations are a natural evolution of references and letters of recommendation. However, they often are more credible than these traditional documents because it is harder to fake a recommendation on LinkedIn than it is to forge a letter. Since many companies are restricting reference checks to verification of title and dates of employment, a LinkedIn recommendation from a supervisor—and/or coworkers—carries weight.

Many describe LinkedIn as a "reputation engine." That's an apt description because your reputation does precede you online—not just in your work history, but also in your LinkedIn recommendations.

Someone looking at your recommendations wants to know two things:

- What are you like?

- Are you good at what you do?

In addition, you can enhance your own reputation by providing recommendations because people viewing your profile can see (and read) the recommendations you make. (Go to the person's profile on LinkedIn, and on the right-hand side of the page, you'll see a box for "(Name) Recommends.") You can see excerpts of their recommendations, or click the link for **See all recommendations.**

Recommendations provide Search Engine Optimization (SEO) results—meaning, they help you get noticed—both on LinkedIn, as well as search engines. Use industry-specific terminology in your recommendations. Keywords included in LinkedIn recommendations also receive emphasis in search engine results—especially searches within LinkedIn.

When conducting a keyword search, all keywords in a profile are indexed, and profiles with a high match of relevant keywords rank higher in search listings. Although LinkedIn's specific algorithms are secret, some experts suggest that keywords in recommendations receive double the rankings of keywords provided in the profile itself.

Building Your Recommendation Base

Recommendations are visible to your personal network and Fortune 500 companies utilizing the LinkedIn Recruiter Tool. They are important primarily because of the perception that they are very difficult to falsify. These recommendations could be one of the factors that land you the job.

How many recommendations you should have on your profile depends on how many contacts you have. A good guideline is 1–2 recommendations for every 50 connections. Ideally, these will be a variety of individuals—not just supervisors, but co-workers, people you supervise, and clients/customers. Choose quality over quantity.

Plan to build your recommendations over time. Because recommendations have a date attached to them, don't try to solicit all of your recommendations at once. Don't write and send your recommendations all at once either. Recommendations are date-stamped, so the reader will be able to see when she or he was added to your page. It's best if they are added over time.

All recommendations now fall under your Experience section. The first two recommendations appear by default. Additional recommendations are available for viewing by clicking **five more recommendations** (note the number will change).

▾ 2 recommendations

... View↓

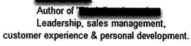

Author of "▮▮▮▮▮▮▮▮▮▮▮
Leadership, sales management, customer experience & personal development.

": ▮▮▮▮ is a tremendously creative resume writer, but her talents exceed that scope. She has the ability to understand... View↓

In the old profile, the names of the people giving the recommendations were all that could be seen without clicking to view them. Now, a good review invites the action of clicking to see what others have said.

The Recommendations Process

The simplest way to get recommendations is to ask. To

request a recommendation through LinkedIn, click the down arrow next to **Edit** when viewing your profile.

You'll find yourself on a page where LinkedIn gives you a summary of recommendations you have received.

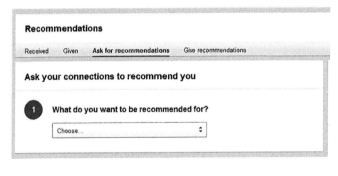

This report shows all the employers and education you have listed in your profile. It also shows you the number of recommendations you have from each employer and educational facility.

Notice that you can ask for recommendations directly from this page. Just click **Ask to be recommended**.

By clicking on the LinkedIn icon, you can add connections to receive your request. Select connections carefully so they are people who can recommend you for the employment listed under "What do you want to be recommended for?"

1 **What do you want to be recommended for?**

Executive Coach/Consulting Campaign Director a ⇕

2 **Who do you want to ask?**

Your connections: (You can add up to 3 people)

Fred Coon ☆ stewartcoopercoon.com ✕

3 Fred Coon ☆ stewartcoopercoon.com

What's your relationship?

Choose... ⇕

What was Fred's position at the time?

Choose... ⇕

4 **Write your message**

Subject:

Can you recommend me?

Write a personable message here|

LinkedIn provides a generic message. It is recommend that you personalize the message. For example, you could select LinkedIn connections that worked with you on a specific type of project. Then ask that set of connections to endorse your work for that project.

Although LinkedIn gives you the option of sending

bulk recommendation requests, don't do it. Each request should be personalized to the individual you are asking for a recommendation, unless there is a common theme that connects all the individuals you wish to connect with.

For example, if you worked with a number of people on a common project, you could send out one message that would be meaningful to all of the individuals you seek to reach.

For example:

"Could you provide me with a recommendation based on our work together on [X Project]?"

Your sample request might look like this:

"Could you provide me with a recommendation based on our work together on your resume? I am developing my LinkedIn profile for this service, and your feedback on the quality of my work would be so helpful!

Thanks!

Susan

Another alternative:

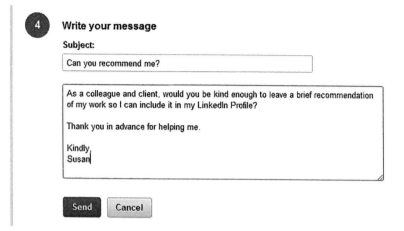

It's even better to send each request out individually. "Hi <name>" is a lot more personal and more likely to receive a response.

Ultimately, the best strategy is to ask for the recommendation through more personal means — for example, in person, on the telephone, or via email.

In fact, one of the best ways to get a LinkedIn recommendation is to ask after they've given you a compliment in real life. If they praise you via email, for example, you could respond with a message that thanks them and says, "Are you on LinkedIn? Would you mind if I sent you a LinkedIn request for a recommendation? It would mean a lot to me to have you say that in a recommendation on there."

Reciprocation is also a powerful motivation for recommendations. Generally, if you ask someone for a recommendation, she or he will expect you to write one for them. (So it's a good idea to only ask for recommendations from someone you'd be willing to write one for in return!) The reverse is also true—sometimes, if you provide an unsolicited recommendation, the person you recommend will go ahead and write one for you, as well.

However, reciprocal recommendations (I gave you one, so can you give me one?) are less powerful than recommendations that are freely given. Remember, visitors to your LinkedIn profile can see who you have recommended as well as who has recommended you. It's easy to spot one-to-one (reciprocal) recommendations.

If you don't receive a response back from someone after requesting a recommendation—or, if you don't feel comfortable following up, consider whether you should be asking for a recommendation from that person in the first place.

One of the most effective ways to get a great LinkedIn recommendation is to write it yourself. This makes it easier on the person who you want to recommend you—and ensures your recommendation is specific and detailed.

In this case, your request for a recommendation might follow this format:

Dear (Name):

I'm writing to request a recommendation of our work together at (company name) that I can include on my LinkedIn profile. To make this easy for you, here's a draft recommendation. Feel free to edit this or create your own.

Thank you.

(Your Name)

When possible, give the person you're asking for a recommendation some context for your request:

"I'm writing to request a recommendation on LinkedIn. As you know, I'm looking to make a career change, and I believe a recommendation from you based on our work together on [X Project] would be useful in highlighting my transferable skills."

How to Handle Recommendations

You'll receive a notification when someone recommends you. LinkedIn sends the notification to the email address on file:

The link at the bottom of the email takes you to the same message in your LinkedIn account (you may need to sign in to your LinkedIn account). It will ask you if you want to "Show this recommendation on my profile" or "Hide this recommendation on my profile." Choose one option and then click **Accept recommendation**.

After you click **Accept recommendation**, you'll receive a **Recommendation Confirmation**. This screen will give you the opportunity to write a reciprocal recommendation.

Checkmark indicates a particular recommendation is displayed on your profile

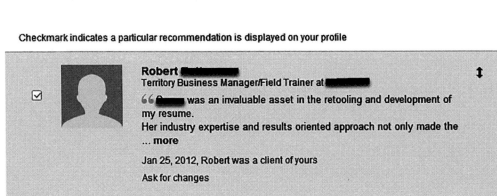

If you find an error in your recommendation, or it's not specific enough, you can click the **Request Replacement** link and it will automatically generate a request for a change with an email to the individual who wrote the recommendation.

STRATEGY TIP

Asking for changes in a recommendation should be handled carefully. Here are some tips that may prevent misunderstandings:

- Correct any spelling and/or punctuation errors. Ask if the corrected version could be used.

- Copy some quotes from your contact's emails. Ask if this information could be included in the recommendation.

- Ask your contact to share what he/she liked the most about working with you and what he/she would want others to know about you.

The best way to handle a recommendation that you don't like is simply to ask for it to be changed. Instead of asking them to change the whole thing, address specific issues in the recommendation that you would like changed.

"I like what you've written, but I was wondering if you would correct the statement where you said I brought in $200,000 in revenue; my records from that time show that the figure was closer to $375,000."

Replace the standard text in the message with your custom message.

Removing Recommendations

You can also choose to remove recommendations from your profile, even after it is published. This is how to manage the recommendations already on your LinkedIn profile: Click **Profile** and scroll down to the **Recommendations** section.

You will see this screenshot:

Recommendations

Ask to be recommended   Manage

Below the screenshot, LinkedIn displays each recom-mendation. By clicking **Manage,** LinkedIn will take you into edit mode where you can edit each one.

Your **Privacy & Settings** link controls whether or not you would like to show or hide your recommendations. Go to the upper right-hand corner (near your mini photo) and in the pull-down menu, select **Privacy & Settings**. Go to the **Manage your recommen-dations** link (see screenshot below).

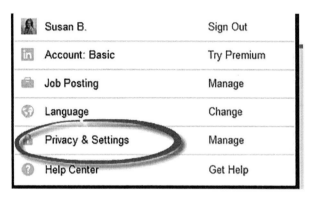

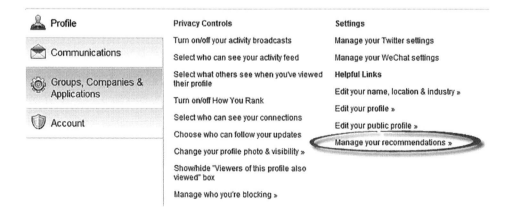

You will see a check box to the left of each recommendation. Checking this box will ensure that your recommendations are displayed. You can also request a new or revised recommendation on this page. For example, if you and a connection worked on a project and would like that recommendation to reflect this effort, then it is worth revising.

What to Do with Poor Recommendations

You can also refuse recommendations. When you receive a message notifying you of a recommendation, if it is not a good one, choose **Hide** this recommendation on my profile.

There are several reasons you might wish to do this. Perhaps whomever wrote recommendation isn't a strong writer. If you prefer not to **Request Replacement**, you could edit the recommendation for grammar and punctuation and ask your connection to update their recommendation with your revised text. In addition, you have the option to Hide the recommendation.

Outside of not accepting a poor recommendation, you can use the steps above to hide it. Simply go to **Manage your recommendations** and uncheck the box to hide the visibility of the recommendation.

It is preferable if you have a good working relationship to use the **Request Replacement** process. You may have coworkers who have weak writing skills, yet would contribute value to your profile with their rec-ommendation if they receive a bit of help from you.

Leaving Recommendations

If you provide recommendations (especially for previous and current colleagues), they are likely to reciprocate, so spend some time crafting thoughtful, unique recom-mendations for people you know.

Formula for Writing LinkedIn Recommendations

Before you write anything, look at your contact's Linke-dIn profile. Align your recommendation with the indi-vidual's LinkedIn profile. Tie in what you write with their headline, summary, and/or experience—reinforce the qualities they want to emphasize in the recommenda-tion you write.

Look at the existing recommendations they've received, as well.

Some things to consider using in your recommendation include:

- What are they good at doing?
- What did they do better than anyone else did?
- What impact did they have on me? (How did they make my life better/easier?)
- What made them stand out?
- Is there a specific result they delivered in this posi-tion?
- What surprised you about the individual?

Choose the qualities you want to emphasize in the person you are recommending.

You may choose to use what author and speaker Lisa B. Marshall calls "The Rule of Threes." Simply stated, concepts or ideas presented in groups of three are more interesting, more enjoyable, and more memorable.

In general, you will want to showcase transferable skills, because these will be the most relevant for your contacts when they are using LinkedIn for a job search or business development.

The top 10 skills employers look for in employees:

- Communication Skills (verbal and written)

- Integrity and Honesty

- Teamwork Skills (works well with others)

- Interpersonal Skills (relates well to others)

- Motivation/Initiative

- Strong Work Ethic

- Analytical Skills

- Flexibility and Adaptability

- Computer Skills

- Organizational Skills

The following formula for a LinkedIn recommendation will help you write a great recommendation.

- Start with how you know the person (1 sentence). Give context for the relationship beyond just the job title and organization/company/school, although that can be a good way to start your recommendation. ("I've known Amy for 10 years, ever since I joined XYZ Company. She was my lead project manager when I was an analyst.")

- Be specific about why you are recommending the individual (1 sentence). What qualities make him or her most valuable? Emphasize what the person did that set him or her apart. What is his work style? Does she have a defining characteristic? To be effective, recommendations should focus on specific qualifications.

- Tell a story (3–5 sentences). Back up your recommendation with a specific example. Your recommendation should demonstrate that you know the person well—so tell a story that only you could tell. Moreover, provide social proof in the story—give scope and scale for the accomplishments. Don't just say the individual you're recommending led the team—say he led a 5-person team or a 22-person team. Supporting evidence—numbers, percentages, and dollar figures—lends detail and credibility to your story.

- End with a call to action (1 sentence). Finish with the statement, "I recommend (name)" and the reason why you would recommend him or her.

In the first sentence, you describe how you know the individual and give context about why you are qualified to recommend him or her.

- (Name) and I have worked together...

- I've known (name) for (how long)...

For the second bullet point, you can set up the description of his or her qualities by providing an overview sentence.

Here are some examples:

- Able to delegate...
- Able to implement...
- Able to plan...
- Able to train...
- Consistent record of ...
- Customer-centered leader...
- Effective in _____
- Experienced professional in the _____ industry
- Held key role in _____
- Highly organized and effective...
- High-tech achiever recognized for...
- Proficient in managing multiple priorities and projects...
- Recognized and appreciated by...
- Served as a liaison between _____
- Strong project manager with...
- Subject-matter expert in _____
- Team player with...
- Technically proficient in _____
- Thrived in an...
- Valued by clients and colleagues for...
- Well-versed in the...

For example:

Mike had a consistent record of delivering year-over-year sales revenue increases while also ensuring top-notch customer service, working effectively with the entire 7-member sales team to make sure the client's needs were met.

Jill is a subject-matter expert in logistics, warehouse planning, and team leadership. Her ability to take the initiative to ensure the prioritization of thousands of items in each shipment for same-day processing made her an indispensable member of the management team.

For the storytelling section, you can choose a Challenge-Action-Result format to describe the project:

> **Challenge:** What was the context for the work situation on the project? What was the problem that the project was designed to tackle?

> **Action:** What did the person you're recommending do? What was their specific contribution?

> **Result:** What was the outcome of the project—and can you quantify it?

Choose descriptive adjectives to include in your recommendations. Instead of describing someone as "innovative," choose a word like "forward-thinking" or "pioneering."

Here are some other descriptive words to consider: For example:

Accessible	Culturally-sensitive	Friendly
Accomplished	Curious	Fun-loving
Accurate	Customer-focused	Funny
Ace	Customer-oriented	Future-oriented
Achievement-oriented	Daring	Generous
Action-driven	Deadline-oriented	Genuine
Active	Decisive	Gifted
Adaptable	Dependable	Global
Adept	Detail-minded	Goal-oriented
Adventurous	Detail-oriented	Happy-go-lucky
Aggressive	Determined	Hardworking
Ambitious	Devoted	Health-conscious
Analytical	Diligent	Healthy
Articulate	Diplomatic	Helpful
Assertive	Directed	Heroic
Authentic	Discreet	High-energy
Authoritative	Dramatic	High-impact
Award-winning	Driven	High-potential
Bilingual	Dynamic	Honest
Bold	Eager	Humorous
Bright	Earnest	Imaginative
Budget-driven	Easygoing	Impressive
Calm	Effective	Incomparable
Capable	Efficient	Independent
Caring	Eloquent	Industrious
Charming	Employee-focused	Influential
Cheerful	Empowered	Ingenious
Collaborative	Encouraging	Innovative
Colorful	Energetic	Insightful
Committed	Enterprising	Inspiring
Communicative	Entertaining	Intelligent
Community oriented	Enthusiastic	Intense
Competitive	Entrepreneurial	Intuitive
Computer-savvy	Ethical	Inventive
Confident	Exceptional	Judicious
Congenial	Experienced	Kind
Connected	Expert	Knowledgeable
Conscientious	Expressive	Likable
Conservative	Extroverted	Logical
Convincing	Fair	Loyal
Cooperative	Flexible	Market-driven
Courageous	Forceful	Masterful
Creative	Formal	Mature
Credible	Forward-thinking	Methodical

Meticulous	Punctual	Team-oriented
Modern	Quality-driven	Team player
Moral	Quick-thinking	Technical
Motivated	Quirky	Tenacious
Multilingual	Reactive	Thorough
Multitalented	Refined	Tolerant
Notable	Reliable	Top-performer
Noteworthy	Reputable	Top-performing
Objective	Resilient	Top producing
Observant	Resourceful	Tough
Open-minded	Respected	Tough-minded
Optimistic	Responsible	Traditional
Orderly	Results-driven	Trained
Original	Results-oriented	Trend-setting
Organized	Rigorous	Troubleshooter
Outgoing	Risk-taking	Trusted
Outstanding	Safety-conscious	Trustworthy
Passionate	Savvy	Undaunted
Patient	Seasoned	Understanding
People-Oriented	Self-accountable	Unrelenting
Perceptive	Self-confident	Upbeat
Perfectionist	Self-directed	Valiant
Performance-driven	Self-driven	Valuable
Persevering	Self-managing	Vaunted
Persistent	Self-motivated	Versatile
Personable	Self-starting	Veteran
Persuasive	Sensible	Visionary
Philanthropic	Sensitive	Vital
Pioneering	Service-oriented	Warm
Poised	Sharp	Well-organized
Polished	Sincere	Well-versed
Popular	Skilled	Willing
Positive	Skillful	Winning
Practical	Sophisticated	Wise
Pragmatic	Spirited	Witty
Precise	Spiritual	Worldly
Professional	Steady	Youthful
Proficient	Strategic	Zealous
Progressive	Strong	
Prolific	Successful	
Prominent	Supportive	
Prompt	Tactful	
Proven	Talented	
Prudent	Task-driven	

Make sure the recommendation you write is clearly about the person you're recommending. That sounds like common sense, but many recommendations are too vague or too general—they could be about anyone, not this specific individual. To be effective, the recommendation you write should not be applicable to anyone else.

Recommendations that you write should be:

- Genuine

- Specific

- Descriptive (with detailed characteristics)

- Powerful (including specific achievements, when possible)

- Honest/Truthful (credibility is important; avoid puffery or exaggeration)

Length is an important consideration when writing LinkedIn recommendations. Keep your recommendations under 200 words whenever possible. Some of the most effective LinkedIn recommendations are only 50–100 words.

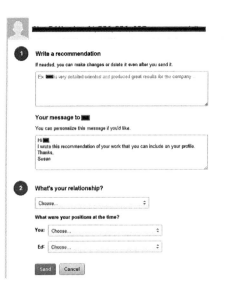

You may find it useful to look at other recommendations before writing yours. You can do a search on LinkedIn for others with that job title and check out the recommendations on their profiles.

You can use LinkedIn's Advanced People Search function to conduct a search. At the top of the page, click the

Advanced link next to the search box.

You can enter in keywords or job titles to find profiles related to the type of recommendation you are writing.

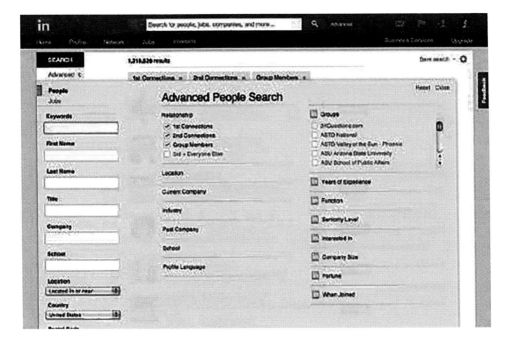

You can then browse the listings that come up as matches and check out the recommendations on those profiles.

Consider drafting your recommendation in Microsoft Word or a text editor. Because LinkedIn does not have a built-in spell-check function, this will help ensure your text does not contain spelling errors. You can also check your grammar in Microsoft Word, and use the Word Count feature to determine the length of your recommendation.

Now you're ready to actually create a recommendation using LinkedIn.

The easiest way to do this is to go back to **view your profile** and hover over the down arrow next to **Edit**. Click **Ask to be Recommended**. This will take you to your recommendations page. Click **Given** on the top left and scroll down until you see **Make a Recommendation**.

You must either be connected to the individual you wish to Recommend or know his or her email address. Also, the individual must have a valid LinkedIn account. You may find it easiest to use the **select from your connections list** in the **Make a Recommendation** section. You can also make a recommendation from the individual's profile page directly.

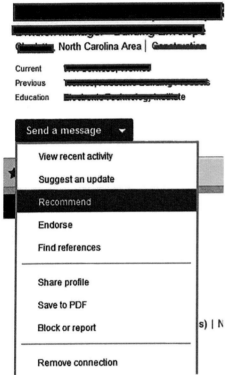

North Carolina Area |

Current
Previous
Education

Send a message ▼

View recent activity

Suggest an update

Recommend

Endorse

Find references

Share profile

Save to PDF

Block or report

Remove connection

On the profile, scroll down until you see **Recommend** in the person's profile. Once you click this, the form pictured on the right will appear.

The Recommend feature may also appear under the **Suggest connections** button. Or the Recommend option might be found in the drop-down menu under Send a Message.

It's easy to recommend your connections because LinkedIn places such encouragements all over. Once you've selected an option, click **Go**. LinkedIn will

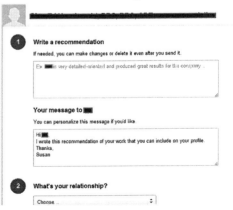

1 Write a recommendation
If needed, you can make changes or delete it even after you send it.

Ex   is very detailed-oriented and produced great results for the company ...

Your message to
You can personalize this message if you'd like.

Hi
I wrote this recommendation of your work that you can include on your profile.
Thanks,
Susan

2 What's your relationship?

Choose ...

guide you through the completion of each step of writing a recommendation.

The person you recommend will get your email notifying him or her that you've made a recommendation. If you don't receive a reply from the individual you've recommended within a week, follow up and make sure they received it.

Recommendations and Endorsements

In 2013, LinkedIn introduced the "endorsement"—a vague and essentially worthless way to tell others what someone in your network knows how to do well. Endorsements make connections feel like they've helped someone in their network. Unfortunately, endorsements don't carry any weight.

Only recommendations carry any real value on LinkedIn. This is one major reason you need to pursue them.

If your request is ignored, there are several strategies to consider.

✓ Make it easy for connections by providing a template. Your connection may not feel comfortable with writing recommendations.

✓ Your connections may have forgotten. A reminder is easy to send. Go to Manage Your Recommendations, click Resend, and edit the message so it's a friendly reminder.

✓ If your connection's memories aren't as positive as you anticipated, withdraw the request.

Keep in mind that you can change (or remove) recommendations you've given. Follow the instructions outlined in the previous page to hide the visibility of a recommendation, if necessary.

You can edit recommendations from this **Manage your recommendations** page and click **Given** at the top. You can scroll to the recommendation you wish to remove and LinkedIn will ask you to confirm the selection.

The connection will receive a notice that you've edited the recommendation.

If you wish to withdraw the recommendation, click **Withdraw this recommendation**. LinkedIn will ask you to **confirm** this change.

Any recommendation you write may show up in your Activity feed on LinkedIn—even before it's approved by the individual you've recommended—so keep that in mind.

Responding to Recommendations Requests

Don't ignore requests for recommendations. But, don't feel as if you must accept all requests to make a recommendation. You can respond back that you don't feel you know him or her well enough to write a recommendation (or that you don't know them well enough in their work life to recommend them, if you only know them socially). Alternatively, you can put them off by saying, "Once we've worked together for a while, I'd be happy to write a recommendation for you."

So-called "character references" (also called "personal references") don't have much of a place on LinkedIn, where the emphasis is on recommendations from people you have worked with ("professional references"). You can say something like, "Although we know each other socially, because LinkedIn attaches recommendations to specific jobs, I don't feel I'm a good fit to write a recommendation for you."

You will rarely see a negative recommendation on LinkedIn. Because the content of recommendations is public, it's likely to be positive, and since recipients can choose to display recommendations or not, they are not likely to approve negative comments for public display.

Your mom was right: "If you can't say something nice, don't say anything at all."

However, if you do decide to write a recommendation, the first question you should ask is "What is the goal?" Does the individual want a new job? A promotion? A career change? A client? Knowing what their goal is in soliciting a recommendation will help you tailor it to meet their needs.

Look at the individual's LinkedIn profile—especially the job description of the position when you worked together.

If asked to provide a recommendation, it is acceptable to ask the person to draft the recommendation in which you can refine.

Remember, recommendations you write show up on your profile too, so someone looking at your profile can see the recommendations you've made for others.

Final Thoughts on Recommendations

Recommendations matter—but who they came from is

sometimes more important than what the recommendation says. A recommendation from a higher-level person makes more of an impact than one from colleagues. You can often judge a recommendation by the quality of the person writing it.

Don't write—or display—bad recommendations on your LinkedIn profile. Bad recommendations are those that are:

- Generic
- From people who don't have a clear understanding of you and/or your work
- Written without context (doesn't include how they know you, how they worked with you)
- Old or outdated

LinkedIn allows you to edit recommendations after posting. However, it is important to remember that you never get a second chance to make a first impression.

Take Action!

On LinkedIn, it's always best to remember that those who give are those that receive. Make "Give to Get" your motto for this week. Take the time to craft 4–6 well thought-out recommendations for your connections this week, and watch for the reciprocating response.

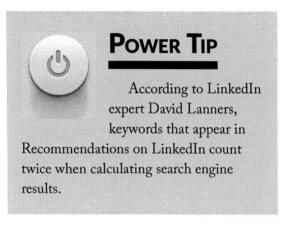

POWER TIP

According to LinkedIn expert David Lanners, keywords that appear in Recommendations on LinkedIn count twice when calculating search engine results.

Finding Jobs Through Company Searches

Finding a job is more than finding an employer who will hire you. It's finding a position that's a good fit for both you and the company for which you end up working. One of LinkedIn's most powerful functions is the insight it provides into various companies and the jobs they offer.

One of the most obvious applications for using LinkedIn in your job search is using the **Job**s tab to identify opportunities. Click **Jobs** in the main navigation bar.

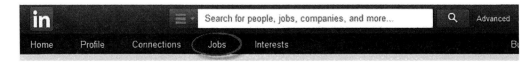

This takes you to the main job search page. By default, the jobs that you'll see on the page are recommendations based upon the keywords you have used in your profile. This is a just a start. You'll want to drill down to find more opportunities.

Note: You may see a message at the bottom of the search results asking you to **Complete your profile to see improved job suggestions.** *If the only obstacle between you and a complete profile is the number of contacts, consider how many requests you have sent out. If you have sent*

out a large number of requests recently, please give peo-
ple a week to respond. LinkedIn will send out reminders
to people you've asked for connections, so you don't have
to become a pest!

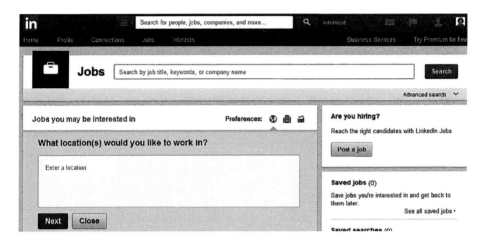

If you enter a specific search in the **Search for Jobs** box
and run a search, LinkedIn takes you to a more advanced
search page. You'll see a sidebar at the left with options
you can select.

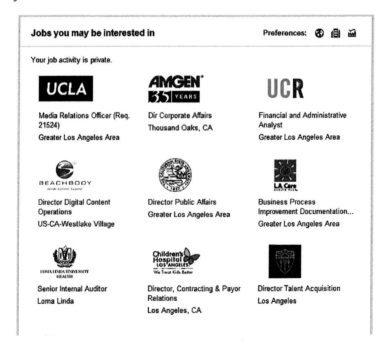

You can identify positions by job title, keywords, or company. If you want to remain in your current location, you can enter the Postal Code. Search results will then fall within your geographic region.

In order to get the broadest results, avoid putting keywords in quotes. If you are only interested in a narrower section of the market, using quotes around your keyword phrase can reduce the number of off-target results.

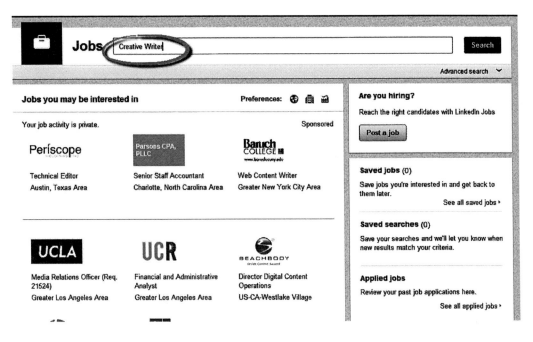

Note: If there aren't any LinkedIn jobs that fit your keywords, LinkedIn will search Simply Hired.

While you are reviewing job matches, you'll notice that you will have the option to **Save** it. If you are interested in the position, click **Save** to add it to your **Saved Jobs** area. After you've

completed your search for the day, you can then visit the **Saved Job**s tab to review all the positions that interest you.

You'll notice that above your search results, there's a **Save Search** in the upper right corner. Click there if you want to create searches that you can save and have emailed to you daily, weekly, or monthly.

If you click the job title, it will take to you to the Job Description. Some jobs allow you to apply for the position directly from the page, whereas other postings will take you to the company's website.

On the right side of this page, you will see **People Also Viewed** with links to other similar jobs.

There's one thing you should keep your eye out for when you see a job posting. You should watch for jobs where someone on LinkedIn is in your network. In the listing below, you can see that there is a 1st tier connection.

People you know at Keller Williams Realty, Inc.

Reach out to your connections for a referr

You

If the person connecting you to the company is a 2nd tier connection, this is the perfect time to find out which 1st tier connection is linking you both.

Click the **1 person** to find out who the 2nd tier connection is. Then click the name to visit the connection's profile. Once there, click the triangle after **Send InMail.**

Select **Get Introduced** in the drop-down menu, and then

follow the steps shared in the last lesson for requesting introductions.

Approximately 50 percent of the jobs will also show you who posted the job.

You can send the job poster an InMail and let them know that you applied for the job and would love to have a conversation about how you are a great fit.

Contact the job poster

Reach out for more information or to follow up on your application.

Experienced Marketing...

Send InMail

Conducting Company Searches

You can also research companies on LinkedIn. Hover your cursor over **Interests** in the main navigation toolbar, and select **Companies** from the drop-down menu.

Note: If there are any companies within your network, they will appear in the drop-down list. Someone you have listed as an employer that has a listing on LinkedIn is typical.

You don't have to know a company name to perform a company search. Use keywords or industry to find companies.

Notice as you scan through the company search results that LinkedIn tells you if any job postings are open. These are the most obvious opportunities during the first round of job-search efforts.

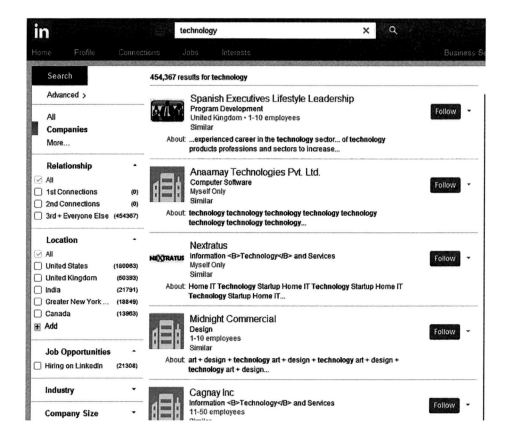

By clicking a company name, you can view all of their details. LinkedIn doesn't give you the option of saving your Company search, so it's a good practice to hold down the **CTRL** key when you click. This will open the Company page in a new window. That way you can explore the Company's LinkedIn information without losing your search.

At the top, just below the company name you'll see four tabs. The most useful from a job search perspective is the **Careers** tab. In addition to finding all job openings there, you may find profiles of employees at the company. Job postings show up at the left side under J**obs at ...**

At the same time, don't overlook the information that

you'll find on the **Home** page. A company blog may be a major resource for you, especially if you do a company search for Resume Writers. For example, one search result had a blog that would give you insights into "Best practices on countering a counter offer." Knowing this information could give you the edge in when negotiating for salary and other benefits.

Products offered by the company give you additional insights into the business. Often these products will have endorsements from LinkedIn members.

Insights looks different from company to company. Some pages will only show you "similar search" results, such as members "who viewed this company, also viewed the following companies." Others provide a lot more information about the business.

This section will allow you to see how many employees are joining or exiting company, the composition of job functions, and whether you have any connections to current employees.

You may also capture a glimpse into how easy it is to move up the ladder in the company. Things like new titles can indicate a move up in the organization.

Unfortunately, some of the more useful statistics from a job search perspective are more difficult to dig for now. Insightful Statistics used to allow you to see how many employees were leaving or coming into the company at a glance.

You can still see what the composition is of job functions within the company and whether you have any connections to current employees at the company. If you see the invitation **View all employees >>** at the upper right side in the people section on the **Careers** page, there are more employees for you to check out.

Another thing you can look for when you are looking at the employee composition of the company is trends. For example, you may notice all the senior-level managers have MBAs or that profile pictures reflect a younger company.

Follow Companies

You can also use the **Follow Company** feature to track prospective employers. When you follow a company, you'll get notices of major changes and notifications when the company loses, gains, or promotes staff (which can be useful to see which companies have a lot of turnover).

To follow a company, first find it using the search feature. Then click on the company's page. On the right-hand side, you'll see a **Follow** button. Click to follow the company.

You'll be able to access a list of the companies you follow from the home page for the **Companies** tab. Just click **Companies** on the main navigation toolbar. Then click the **Following** tab. Uncheck **Following** box to stop following a company.

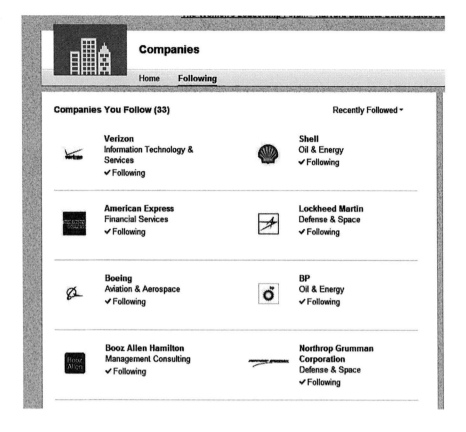

Take Action!

Use the **Job**s tab to identify job openings, but don't stop there. Identify 2–3 companies that you'd like to work for which have openings.

Search those companies on LinkedIn to learn more about them. Follow these companies, and use Insightful Statistics data to connect more fully.

Checklist for Getting the Most Out of Companies

Click the Companies tab to:

- Search for companies directly by name, keyword, or industry.

- Use the Search Companies tab for more advanced search options, refining your search by location, company size, number of followers, and relationship.

- Find companies to follow to stay up to date on conversations and business development related to the company itself and its contacts.

- Review and recommend people in your industry or area of interest that you've had contact with or are interested in contacting.

THE POWER OF GROUPS

Joining a LinkedIn Group provides you with opportunities to strengthen connections with like-minded individuals in an exclusive forum setting. The Groups function provides a private space to interact with LinkedIn members who share common skills, experiences, industry affiliations, and goals. You can easily find groups within your industry to join, as well as local groups.

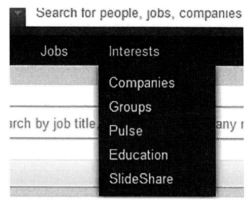

Laura DeCarlo's Critical 3-Part LinkedIn Strategy

✓ Complete your profile in full.

✓ Reach out to colleagues, coworkers, and bosses and give them recommendations before you expect them to be written for you.

✓ Become an ACTIVE member in as many relevant groups for your profession and industry, as possible.

These easy steps will allow you to: a) make contacts in your sphere, b) improve your search-ability, and c) make your presence value-added. Laura DeCarlo, President of Career Directors International – www.careerdirectors.com.

You may view your Groups by clicking **Groups** from the **Interests** tab on the navigation bar or you may search for **Groups** by using keywords or a group name in the search field at the top and clicking **Groups** on the left navigation pane.

Narrow your search with the category and language selections, as necessary. Click the search icon.

There are two types of Groups: Open groups and Members only. A padlock signifies Members only Groups.

Career Advice for Event Planning & Management

a subgroup of **Event Planning & Event Management - the 1st Group for Event...**

Note: Open groups may be viewed and participated in by anyone. Members' only groups are better for developing credibility among peers. Open groups are better for growing general visibility.

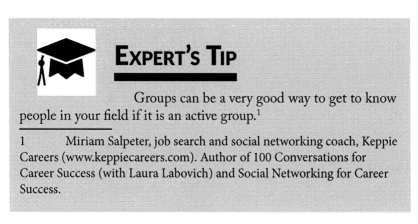

EXPERT'S TIP

Groups can be a very good way to get to know people in your field if it is an active group.[1]

1 Miriam Salpeter, job search and social networking coach, Keppie Careers (www.keppiecareers.com). Author of 100 Conversations for Career Success (with Laura Labovich) and Social Networking for Career Success.

You'll notice that this search produced a list of groups that target someone who wants to develop his or her credibility in the job-search coaching arena. Apply this same principle to your search for groups. Narrow your choices by using industry and career-focused keywords.

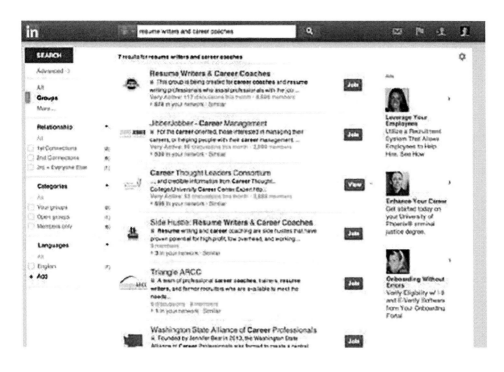

EXPERT'S TIP

LinkedIn Groups is one of the two most important features in LinkedIn. It's critical to understand how to participate and get value out of Groups[1].

1　　　Jason Alba, Author of I'm on LinkedIn – Now What??? and Founder of JibberJobber.com

Groups You May Like

Because this list is so large, LinkedIn also offers Groups you may be interested in joining. You can access it when you view your **Groups.** Recommendations are based on the information you provided in your profile, and they are usually related to your industry.

Discover more

Once you find a group that you would like to join, simply click the **Join** button. If it is a Members only group, you will get a message that your request to join has been received.

Once you are approved, you will receive a notification in your email account. You can then start participating in the group. If you request to join an open group, you will automatically be a member and will receive a welcome message to the group.

Checklist for Maximizing Your Use of Groups

Click on the Groups tab to:

- Review groups you're currently a member of, if any.

- Search for "Groups You May Like" based on primary keywords, categories, languages, or similar groups within the same interests.

- Check the Groups Directory for featured groups in your niche or areas of interest you might like to join.

- Create a Group of your own to attract followers, entering a logo, group name, type, brief and full description, website, and email address as well as access option, language, and location.

If you are joining a group to develop relationships with influencers, participate in the group to gain visibility before you invite them to connect with you on LinkedIn.

To establish a presence, ask questions, give advice, and be helpful to others. You can also have group notifications emailed to you.

Groups with the largest number of members aren't always the best groups in which to participate. Some groups require a high standard from members, and thus the number of members remains lower. Don't let this dissuade you from considering a group. Some of the best rests among members of very small, yet selective groups.

Should I start a group? Or is it better to start a Discussion in a Group?

Answers from the Experts

- If you have the time and energy for a new group, and you have the follow-through to expand and grow the group, it can be a great way to demonstrate leadership in your field. However, if there is already an active group relevant to your expertise, it can useful to become an active participant and influencer.

- It's easiest to start a Discussion in an existing Group that already has an audience. Owning a group is powerful, but most job seekers want immediate results. Starting your own Group will have long-term benefits, but you might not see those benefits as soon as you need them.

Leveraging the Power of Media

Rich Media

LinkedIn offers the ability to add Rich Media. There is a media button in the following three sections: Summary, Education, and Experience.

LinkedIn allows you to add videos, images, documents, and presentations directly to your profile. But first, you need to connect third-party media providers to your account. '

Otherwise, LinkedIn's instructions for how to add rich media will not work. Once you add them, you can rearrange the media through drag-and-drop features.

Media Link Providers

Here are the official third-party applications that LinkedIn recognizes. Create public accounts and then link them to your account. Use them to build your visibility in a job search.

Video Providers

- ABC News
- AllThingsD
- Animoto
- bambuser
- big think
- blip
- Boston
- Brainsonic
- Bravo
- brightcove
- CBS News
- Clikthrough
- Clipfish
- Clip Syndicate
- Clipter
- CNBC
- CNN
- CNN Edition
- CNN Money
- Colbert Nation
- CollegeHumor
- Comedy Central
- Confreaks
- Coub
- Crackle
- Dailymotion

- DeviantART
- distrify
- dotsub
- Fora.tv
- Forbes
- funny or die
- GameTrailers
- GodTube
- Hulu
- Img.ly
- JibJab
- Khan Academy
- KoldCast TV
- LiveLeak
- Logo FierceTV
- photozou
- TED
- Washington Post
- Zapiks
- Telly
- Dribble
- The Escapist
- mobypicture
- New York Magazine

- NZ On Screen
- Overstream
- PBS Video
- Revision3
- SchoolTube
- ScienceStage
- ShowMe
- snotr
- Socialcam
- Spreecast
- VEVO
- Viddler
- viewrz
- Vimeo
- orldstarhiphop
- xkcd
- xtranormal
- Youku
- YouTube

Image Providers

- 23hq
- Mead
- mlkshk
- mobypicture
- ow.ly

- pikchur
- Pinterest
- Questionable Content
- somecards

- twitgoo
- twitpic
- TwitrPix
- Twitter

Audio Providers

- AudioBoo
- Band Camp
- Free Music Archive
- gogoyoko
- Grooveshark
- Hark
- Huffduffer
- Mixcloud
- RadioReddit
- Rdio
- SoundCloud
- Spotify
- Zero

Presentations and Documents

- Google Docs
- Prezi
- Scribd
- lideShare

Other

- Behance
- Issuu
- Kickstarter
- Quantcast

Also, you may upload the following files directly to your profile.

Presentations		Documents		Images	
.pdf	ppsx	.pdf	.rtf	.png	.jpg
.ppt	.pot	.doc	.odt	.gif	.jpeg
.pps	.potx	.docx			
.pptx.	.odp				

As you can see, there are far more media options available

to you now. It might seem that LinkedIn makes changes often, yet often they are very beneficial changes.

Once you have the new profile, you'll be able to add media content from any one of the providers listed above.

Adding content is as simple as going into profile editing mode. Scroll down until you see the section where you would like to add content. Click the media button. (It looks like a rectangle with a plus at the lower right corner.).

All you have to do is type in the URL for whatever type of media you want to add to your profile or upload the file from your computer.

LinkedIn immediately recognizes whether the URL comes from one of the supported providers or not. If it does, you'll be sent a window with prefilled **Title** and **Description** fields.

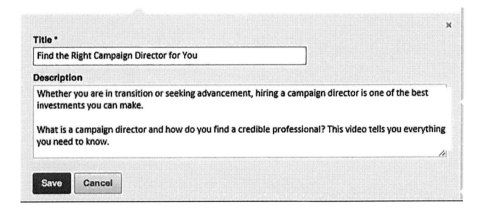

Note: The prefilled title may be ugly and look like a file name. Change that. Give it a good title and rework the description, as necessary.

Be sure you click **Save.**

Take Action

Identify which type of media could help you attain your employment goals. Set a goal to prepare and upload at least one media presentation in the next week.

FINAL THOUGHTS

If you've been taking action, as recommended, you've already:

- Created your LinkedIn account

- Optimized your LinkedIn profile (including photo, headline, and summary)

- Set your LinkedIn privacy settings

- Made connections with people you know

- Researched (and followed) companies you'd like to work for

- Joined a number of LinkedIn Groups

- Requested and received recommendations to complete your profile.

However, to get the most out of LinkedIn — especially for your job search—you need to make LinkedIn a regular part of your routine.

If you're in an active job search, your routine should include logging in to LinkedIn each day to:

- Update your status

- Make new connections

- Research employers

- Read — and start — threads in your LinkedIn Groups

- Give recommendations to your connections

Now, here are some additional strategies you can use to uncover hidden opportunities.

Nurture Local Connections

Search LinkedIn for businesses and potential contacts using a keyword/area code search. Much of your job search strategy focuses on developing a person-to-person network, and LinkedIn is a good place to find out whom you need to connect with in person. Then, you can uncover who you know that can help you meet the right people via LinkedIn.

That's a first step toward meeting in person for an informational interview, a very useful strategy for any job seeker who is finding it difficult to locate openings to which he or she may apply.

Another opportunity you should look into on LinkedIn is volunteer openings at local not-for-profits. Filling a time of unemployment with volunteer activities looks good on your LinkedIn profile. It helps dispel the perception that you are unemployable especially when your job search is within one of the sectors that has experienced a downturn in recent years.

Leverage More Out of Company Information

As you are looking through the information a company has posted on LinkedIn, check out the profiles of the people they have hired. Look to see what their employment history has been. You may discover some useful strategies for your own personal career goals, especially if you are starting out.

For example, you can look for trends in past work history

within certain departments of a company. You might also discover a trend toward hiring employees who work outside of the traditional office environment and use technology to create the work environment instead.

Search Companies by Skill Set

You may have done this to some extent by the keywords you've entered in previous company searches. Now shift your focus. Use an advanced search to discover where the most jobs are located for your skills and expertise. It could be that you will find it much easier to find work if you look outside your local area.

For example, the demand for graphic artists is very depressed in the Portland Metro area. You might find opportunities are opening up in other regions. Alternatively, it might be easier to build your own business if you were living in an area where the cost of living weren't as high, and high-speed internet was available.

Get to Know Hiring Managers

When you search for jobs, pay extra special attention to those jobs where a hiring manager is a 2nd or 3rd tier connection. Someone you know is acquainted with the person who posted the job. This is an opportunity you should not ignore.

If your connection is willing to recommend you, you may be able to set up an information interview in person or over the phone.

If you don't have any direct connections, use LinkedIn to identify who the hiring manager is. To increase your chance of success, direct your application properly.

Another strategy you can use is to find a connection within the company who is willing to walk your application to the hiring manager's desk. HR doesn't have to know the person handing the resume over. Just the fact that the person works within the company can help your resume get attention.

Most companies don't spell out all the things they are really looking for in an applicant. Finding a company connection within your network can help you acquire inside information about a listed position. You might also find out about non-posted opportunities.

Consider Startups

LinkedIn is a place where new companies are seeking exposure. Use an advanced search to locate startup or stealth opportunities. Try both the Company and the Keyword fields, as the results will change. Since job security with large corporations is a thing of the past, consider startup opportunities, if your finances support it.

Check Out LinkedIn Labs

LinkedIn Labs offers a number of applications built by LinkedIn employees. There are a large number of apps focused on veterans, which could be useful for you if you have been in the military. Many are primarily visual, yet others could be useful such as Resume Builder.

To find the most current applications, visit **http://engineering.linkedin.com/linkedinlabs/.**

Remember, LinkedIn is a social site—the more you put into it, the more you will get out of it. Get in the habit of using LinkedIn to research opportunities and make connections with individuals who can help you with your job search.

When you have a job interview lined up, search for the interviewer on LinkedIn. See who you know in common, and research the interviewer's background. Review the company profile as well, and see if you have any connections with current employees.

Stay up to date with the latest happenings on LinkedIn too. Follow their blog at: **http://blog.linkedin.com/** Remember to always give! It will come back to you in some way.

DON'T MAKE THESE LINKEDIN MISTAKES

Don't Dismiss LinkedIn as Something Only for People Who Are Looking for a New Job.

The best time to build your LinkedIn profile, connect with people, and participate on LinkedIn is before you need it. If you find yourself suddenly unemployed and decide that now is the time to start using LinkedIn, you're going to be playing catch up.

Instead, take time to "dig your well before you're thirsty," as author Harvey Mackay says.

Don't "Set it and Forget It."

Your LinkedIn profile is an evolving snapshot of you. You should be updating it regularly with new connections, status updates, and activity (within LinkedIn Groups and LinkedIn Answers, in particular).

Check in on LinkedIn regularly—at least every other day if you are in active job search mode or at least once a week if you are a passive job seeker. Plan to add one new status update each time you log in.

Don't Forget to Explore the People Your Connections Know

One of the most powerful functions of LinkedIn is the ability to connect you with people who are connections of the people you know. Follow LinkedIn's guidelines on connecting with these folks (using InMail or requesting connections through your mutual friend) so that your account is not flagged for spam.

Don't Be a Wallflower

LinkedIn is most effective when you engage with it. Seek out opportunities to connect with thought leaders in your industry. Join 3–5 Groups and participate in conversations. Respond to or ask questions in the LinkedIn Answers section.

Don't Be Selfish

You will get more out of LinkedIn if you focus on how you can help others, not how they can help you. The phrase "give to get" is very powerful on LinkedIn. You can earn the respect of your peers and people of influence if you "help enough other people get what they want," in the words of Zig Ziglar.

Don't Wait for Others to Find You

Use the LinkedIn People Search function to look for people you know and invite them to connect with you. You should aim to add 2–5 new connections each week if you are a passive job seeker and 6–10 connections a week if you are actively searching for a new job.

Don't Forget to Check Out "LinkedIn Today"

On your home page of your LinkedIn profile is a round-up of stories that LinkedIn views as interesting to you. Check out these Top Headlines to stay abreast of important information in your industry.

Don't Forget to Give Recommendations

Acknowledge and recognize the contributions of people you know by providing unsolicited, genuine recommendations for them.

Don't Restrict Your LinkedIn Networking to Online Only

Use LinkedIn to connect with people, but then request in-person get-togethers, when possible. Meet for coffee or lunch to catch up. The LinkedIn Events section can also alert you to in-person gatherings in your industry or geographic area.

Don't Try to Connect with People Indiscriminately

One of the strengths of LinkedIn is the connections you make, but it's not a race to get to 500 connections. Have a reason for each person with whom you connect—either it's someone you already know or related to, or someone beneficial to connect with.

If you don't know someone, take time to get to know the person before sending a personalized connection request. (You can also see if you share any connections in common or by checking out their LinkedIn summary and work history, visiting their website or blog, and viewing which Groups s/he belongs to).

Resources to Make Your Job Search on LinkedIn Easier

Interview Readiness

15-SecondPitch.com

This site will help you trim your elevator speech so you are ready for those calls.

Zoominfo.com

This is a great site for locating companies and doing your background research.

Join.me

This website makes it easy to set up client conferences and screen sharing.

Social Media Management

HootSuite.com

There is a fee to use this service, but if you want to manage social media profiles on several sites, this is a very efficient way to do it.

Tweetdeck.com

This is a free web application that can be installed on your computer. There are versions for iPhone, Android, and Chrome. It will allow you to manage social media connections on LinkedIn, Twitter, Facebook, MySpace, Foursquare, and Google+.

TwitterFeed.com

Use TwitterFeed to set up RSS feeds to your LinkedIn account. Every time you publish a new blog, TwitterFeed will post a notice to your account.

Engaging Profile

Vizualize.me

Use this free service to create an infographic resume from your LinkedIn data.

Tinyurl.com | Bitly.com |Cli.gs

These are three link-shortening services. Cli.gs has great analytics so you can trace the effectiveness of a link.

ABOUT THE AUTHORS

Fred Coon

Fred is the founder of Stewart, Cooper & Coon. He is a Licensed Employment Agent, a Nationally Certified Job and Career Transition Coach, a Behavioral Consultant, and a Certified DISC Administrator and author of multiple job search books. He has advised thousands of executives on their job search campaigns.

Fred authored the best-selling career book titled *Ready Aim Hired*, now in its third printing, and is a contributing author to *Business Model You*, by Tim Clark, which is translated into seven languages. His other books include, *Leveraging Linked In For Job Search Success; It's EQ, Not IQ, Stupid;* and the *MasterClass Series* mini-books on job search.

He is a volunteer training adviser for the United States Marine Corps, Wounded Warrior Battalion, and provides interview training workshops at Marine Corps Air Ground Combat Center at Twentynine Palms, CA.

Fred is a member of the Arizona Technology Council's Policy Advisory Committee and a member of the Quality Standards Committee of the Governor's Workforce Arizona Council. He has been on the executive teams responsible for two different companies achieving multiple Inc. Magazine listings in the top 500 fastest-growing companies in America.

He is also a well-known five-string claw hammer banjo player, with appearances on local, regional, and national radio and television programs for over four decades, with stage appearances in The British Isles, Europe, and Australia.

Susan Barens

Susan is an executive coach and campaign director with more than 25 years of experience leading career management, outplacement, leadership development, and strategic communications. Dedicated to demystifying the search and placement, Susan helps her clients turn their unique experiences, themes, and brand into actionable marketing strategies.

Susan is an active member and contributor to industry-related associations, popular book publications, and mainstream newspapers and websites. She is a credentialed Global Career Development Facilitator (GCDF), Certified Professional Resume Writer (CPRW), International Job & Career Transition Coach (IJCTC), and a Master Federal Career Coach & Trainer (MFCCT). Known as a "difference maker," Susan's clients include Fortune 500 executives, academic professors, and military and government leaders within the Department of Defense, Pentagon, Intelligence Community, and the White House.

Susan is a strong advocate of hospice care and veteran causes. She uses her enjoyment of distance running to burn off the energy that comes from being a history and political science junkie as well as a cooking/baking aficionado.

Sources

Block, Jay. *A Picture's Worth a Thousand Words? FALSE!* "Spotlight." Professional Association of Resume Writers & Career Coaches, March 2015, 7–9.

Hempel, Jessi. Last modified March 24, 2010. **Money.cnn.com/2010/03/24/technology/** linkedin_social_networking.fortune/.

LinkedIn. (2014, August). LinkedIn Help Center. Retrieved from **https://help.linkedin.com/app/answers/detail/ a_id/5023/kw/outlook+contacts+import/related.**

LinkedIn Talent Solutions. *Build Your Personal Brand on LinkedIn.* Last modified February 2015. **https://business.linkedin.com/talentsolutions/c/14/ 12/build-your-personal-brand-on-linkedin-g.**

Wynkoop, Kooper. *The Importance of LinkedIn During a Job Search | Jobfully Blog.* Jobfully Blog. Free Resources for Job Seekers to Help Them Find a Job Faster and Easier. Accessed March 9, 2015. **http://blog.jobfully. com/2012/05/linkedin-and-job-search/.**

INDEX